S0-CDP-531

Artificial Intelligence

Springer
Berlin
Heidelberg
New York
Barcelona
Budapest
Hong Kong
London
Milan
Paris
Santa Clara
Singapore
Tokyo

Qiang Yang

Intelligent Planning

A Decomposition and Abstraction Based Approach

Foreword by Martha Pollack

Professor Qiang Yang
Simon Fraser University
School of Computing Science
Ebco/Epic NSERC Industrial Chair
Burnaby, British Columbia
Canada V5A 1S6

With 76 Figures and 49 Tables

ISBN 3-540-61901-1 Springer-Verlag Berlin Heidelberg New York

Library of Congress Cataloging-in-Publication Data
Yang, Qiang, 1961– Intelligent Planning: a decomposition and abstraction based approach / Qiang Yang; foreword by Martha Pollack. p. cm. – (Artificial intelligence)
Includes bibliographical references and index.
ISBN 3-540-61901-1 (hardcover: alk. paper).
1. Artificial intelligence – Congresses. 2. Planning – Data processing congresses. I. Title. II. Series: Artificial intelligence (Berlin, Germany)
Q334.Y36 1997 006.3–dc21 96-29507 CIP

Cover design: Künkel+Lopka, Heidelberg
Typesetting: Camera-ready by the author
SPIN 10553885 45/3142 - 5 4 3 2 1 0 - Printed on acid-free paper

Foreword

"The central fact is that we are planning agents."
(M. Bratman, *Intentions, Plans, and Practical Reasoning*, 1987, p. 2)

Recent arguments to the contrary notwithstanding, it seems to be the case that people—the best exemplars of general intelligence that we have to date—do a lot of planning. It is therefore not surprising that modeling the planning process has always been a central part of the Artificial Intelligence enterprise. Reasonable behavior in complex environments requires the ability to consider what actions one should take, in order to achieve (some of) what one wants—and that, in a nutshell, is what AI planning systems attempt to do.

Indeed, the basic description of a plan generation algorithm has remained constant for nearly three decades: given a desciption of an initial state I, a goal state G, and a set of action types, find a sequence S of instantiated actions such that when S is executed in state I, G is guaranteed as a result. Working out the details of this class of algorithms, and making the elaborations necessary for them to be effective in real environments, have proven to be bigger tasks than one might have imagined.

Initially, plan formation was approached as a formal process, in particular, in Green's work on planning as theorem proving [59]. But beginning in the early 1970s, the focus of the planning community shifted to system development, starting with the STRIPS planning system, which was used, amongst other things, to enable Shakey-the-Robot to form plans to push blocks around the halls of SRI International. STRIPS was followed by a series of ever larger, more complex, and, alas, often more *ad hoc* planning systems. A major break occurred in the late 1980s, marked by the publication of three key papers: Chapman's paper on the TWEAK formalism [28], Pednault's paper on the ADL formalism [104], and McAllester and Rosenblitt's paper on the SNLP algorithm [94]. These papers were intended not to add functionality to known planning methods, but rather to capture the essential elements of these known methods in a readily analyzable fashion. As such, they signaled the beginning of an effort, still ongoing within the planning community, to address the planning problem more systematically, giving greater care to analyzing the relationships among alternative representations and algorithms.

With this shift, the field of planning was ready for a comprehensive book on planning that gives the "lay of the land", showing what methods exist, how they are related, and what their relative strengths and weaknesses are. *Intelligent Planning: A Decomposition and Abstraction Based Approach* is that book. General AI textbooks can typically devote only a chapter or two to the topic of planning, and so can give only a suggestion of the range of issues that arise in plan generation, and of the range of solutions that have been proposed in response. In contrast, this book explores the issues and the solutions in depth, giving a careful and thorough analysis of each. It takes the perspective that effective planning relies on two techniques that are fundamental in computer science—decomposition (also often known as divide-and-conquer) and hierarchical abstraction—and it uses this perspective to structure the material very effectively.

Comprehensive monographs are already available for several other subareas of AI, such as natural-language processing [4] and machine learning [86]. These books have played an important role, bringing together the major ideas of their respective areas to provide a solid platform on which further research can be based. *Intelligent Planning: A Decomposition and Abstraction Based Approach* will do the same for AI planning. As a textbook, it will doubtless be a valuable resource for graduate students and their professors. It will also be a valuable resource for researchers actively working in the field of AI planning, and those in other areas who need to know about AI planning, as it provides ready access to the basic computational tools—representations, algorithms, and analyses—on which further research into the nature of planning will rely.

Martha Pollack

Preface

Those who triumph
Plan at their headquarters
Considering many conditions
Prior to a challenge.

Those who are defeated
Plan at their headquarters
Considering few conditions
Prior to a challenge.

Much planning brings triumph
Little planning brings defeat
How much more so
With no planning at all!

Observing a planning process
I can see triumph and defeat.

Sun Tzu [400-320 B.C., China], *The Art of Strategy*[1], Chapter One

Planning has captivated human interest for generations. The above quote, translated from a classical Chinese text on the art of strategy, underscores this fascination. To live our lives, we have to deal with a huge number of problems, many of which require careful planning. However, until recently the problem of *how* to plan has not been a subject of systematic study. This situation, however, has completely changed with the dramatic progress of computer technology and the enormous success of Artificial Intelligence (AI).

This book is a monograph on Artificial Intelligence Planning (AI Planning), an active research and applications field for the past several decades. As a research field, planning can be defined broadly as the study of actions and changes, covering topics concerning action and plan representation, plan synthesis and reasoning, analysis of planning algorithms, plan execution and monitoring, plan reuse and learning. Lately there has been a dramatic increase of interest in automatic and semi-automatic planning in AI and other related fields. It has been demonstrated that research in planning is of great importance to most subfields of AI as well as general computer science, engineering, management science, and cognitive science.

---

[1] The quotes in this book are adapted from an excellent translation by R.L. Wing [137]. The book is also known as *The Art of War*.

Features

To cover the entire field in a single volume is impossible. In writing the book, I have chosen to focus on a clear, thorough coverage of key areas of classical AI planning. Classical AI planning is concerned mainly with the generation of plans to achieve a set of pre-defined goals in situations where most relevant conditions in the outside world are known, and where the plan's success is not affected by changes in the outside world. As we will see, the planning task is extremely hard even under these situations.

The book's main purpose is to build more intelligence on a set of basic methods for reasoning about and generating plans. This is done by first explaining these basic methods, then developing advanced techniques to enhance them using *problem decomposition* and *abstraction*. In addition, the book presents techniques for analyzing and comparing planning algorithms.

In order to be accessible to readers from a wide variety of backgrounds, the book takes a ground-zero approach. It begins with a gentle introduction and is self-contained; most key algorithms and techniques can be found in the book itself, rather than referenced from other sources. As a result, it requires minimal preconditions on the part of the reader and will benefit not only seasoned researchers, but undergraduate and graduate students, as well as researchers in other related fields such as mechanical engineering, business administration and software project management. Many useful algorithms and techniques are compiled in a single volume, expressed in a common syntactic framework. Many illustrations, examples, algorithms, analyses, tables, and references help make the explanation clear. Each chapter ends with a background survey of the current state of research, the source of the material under discussion, and an exploration of open problems.

Contents and Intended Audience

The book consists of three main parts. Part I, "Representation, Basic Algorithms and Analytical Techniques," lays the foundation. This part provides a general introduction to plan representation, generation, and analysis. It reviews past and current representations and algorithms in AI planning and general computer science that are basic but foundational.

Parts II and III of the book present my own contribution to planning in the past decade. Both parts develop advanced planning techniques that are based on the basic planning algorithms and methods from Part I. Part II, "Problem Decomposition and Solution Combination," presents a complete suite of analysis and algorithm tools for decomposition planning. Decomposition planning refers to the task of breaking apart a complex problem into smaller pieces and solving them individually, and later combining their solutions into a global solution.

Part III is entitled "Hierarchical Abstraction," and presents a theory of plan generation using the idea of abstraction. Abstraction refers to the task of solving a problem by tackling the most important components first, then using the "skeleton" solution as a guide to solve the rest of the problem. This part presents analysis, comparisons between planning with and without abstraction, methods for generating an abstraction hierarchy automatically, properties of hierarchical task network (HTN) planning, and effect abstraction.

This book can be used as a one-semester or one-quarter textbook in an Introduction to AI course, or an AI Planning course. It can also be used as a reference book for graduate seminar courses. As a teaching resource, it can be used in both graduate or undergraduate courses. As a reference resource, it can be used by researchers and practitioners in AI, computer science, engineering, and other related fields. Knowledge of basic data structures, logic, and algorithm analysis would be helpful, but is not strictly required. Any programming experience would also be very useful.

Acknowledgment

The book surveys the field of planning and reports work resulting from a long and enjoyable collaboration with many colleagues and students. Many people helped enrich my view of planning during my graduate studies at the University of Maryland, in the USA, and during my tenure at the University of Waterloo and Simon Fraser University in Canada. I would like to thank my former supervisors Dana S. Nau and Jim Hendler, who sparked my interest in the field of planning and collaborated with me on many challenging problems. Thanks also to my former students at Waterloo whose talent and enthusiasm motivated me throughout my research and teaching career: Steven Woods, Eugene Fink, Philip Fong, Alex Chan, Cheryl Murray, Stephanie Ellis, and to my research associates at Simon Fraser University Edward Kim and Toby Donaldson. Among my close colleagues, I'd like to express my special thanks to Ming Li, Fahiem Bacchus, Josh Tenenberg, and Craig Knoblock. Specific mention of collaborative projects with my colleagues can be found at the end of each chapter.

The book project received strong encouragement from Martha Pollack, a prominent researcher in AI planning and winner of the prestigious *Computer and Thought Award.* Martha read through the manuscript carefully despite her busy schedule, providing many suggestions. Russ Greiner and Henry Kautz gave many in-depth comments on selected chapters. Diane Wudel provided professional editing for the entire manuscript. Thanks also to my former supervisor Dana Nau for introducing me to the Springer-Verlag series.

I would also like to express my deepest appreciation for the support and understanding of my parents Haishou Yang and Xiuying Li, and my wife Jill and son Xin Xin (Andrew). Thank you Jill for tolerating my absence during many evenings and weekends, and for all the late suppers. Without the support and encouragement from my family the project would never have reached completion.

Finally, the work would not have been possible without the strong support from the Natural Sciences and Engineering Research Council of Canada, individual research grant OGP0089686, and support from the University of Waterloo and Simon Fraser University. The book is formatted using LaTeX.

May 1997 Qiang Yang

Table of Contents

1. **Introduction** 1
   1.1 The Problem 1
   1.2 Key Issues 1
   1.3 Planning Versus Scheduling 4
   1.4 Contributions and Organization 6
   1.5 Background 8

Part I. Representation, Basic Algorithms, and Analytical Techniques

2. **Representation and Basic Algorithms** 15
   2.1 Basic Representation 15
      2.1.1 Domain Description 15
      2.1.2 Partial-Order and Partial-Instantiation Plans 17
   2.2 Basic Planning Algorithms 19
   2.3 Partial-Order, Backward-Chaining Algorithms 20
      2.3.1 Correctness 21
      2.3.2 Threat Detection and Resolution 23
      2.3.3 Establishing Preconditions 23
      2.3.4 A House-Painting Example 25
   2.4 Total-Order, Forward-Chaining Algorithms 27
   2.5 More Advanced Operator Representations 29
   2.6 Search Control 31
   2.7 Satisfaction Based Methods 32
      2.7.1 Overview 32
      2.7.2 Model Definition 33
      2.7.3 Model Satisfaction and Plan Construction 35
   2.8 Background 37

3. **Analytical Techniques** 39
   3.1 Basic Analytical Techniques 39
   3.2 Analyzing Forward-Chaining, Total-Order Planners 40
   3.3 Analyzing Partial-Order Planners 41

3.4 Case Study: An Analysis of TWEAK 44
3.4.1 An Implementation of TWEAK 44
3.4.2 Analyzing TWEAK 46
3.5 Critics of Basic Planners 46
3.6 Background 47

4. Useful Supporting Algorithms 49
4.1 Propositional Unification Algorithm 49
4.2 Graph Algorithms 51
4.3 Dynamic Programming 54
4.4 Branch and Bound 56
4.5 Constraint Satisfaction 58
4.5.1 Local-Consistency Based Methods 59
4.5.2 Backtrack-Free Algorithm 60
4.5.3 Backtrack-Based Algorithms 61
4.6 The GSAT Algorithm 64
4.7 Background 66

5. Case Study: Collective Resource Reasoning 67
5.1 Eager Variable-Binding Commitments 67
5.2 Extending the Representation 69
5.3 Plan Generation 71
5.3.1 Correctness Check 71
5.3.2 Threat Detection 71
5.3.3 Precondition Establishment 72
5.4 A House-Painting Example 72
5.5 A Variable-Sized Blocks World Domain 76
5.6 Summary 78
5.7 Background 78

Part II. Problem Decomposition and Solution Combination

6. Planning by Decomposition 85
6.1 Decomposition Planning 85
6.1.1 Two Examples 85
6.1.2 Global Planning-Domain Decomposition 86
6.2 Efficiency Benefits of Decomposition 87
6.2.1 Forward-Chaining, Total-Order Planners 87
6.2.2 Backward-Chaining, Partial-Order Planners 90
6.2.3 Criteria for Good Decomposition 92
6.3 Goal-Directed Decomposition: The GDECOMP Algorithm 92
6.4 Other Benefits of Decomposition 95
6.5 Alternative Approaches to Decomposition Planning 95

6.6 Summary ... 99
6.7 Background ... 100

7. Global Conflict Resolution ... 101
7.1 Global Conflict Resolution — An Overview ... 101
7.2 Conflict Resolution Constraints ... 102
7.2.1 Conflicts and Conflict Resolution Methods ... 102
7.2.2 Relations Among Conflict Resolution Methods ... 104
7.3 Conflict Resolution as Constraint Satisfaction ... 106
7.3.1 Representation ... 106
7.3.2 Propagating Constraints Among Conflicts ... 107
7.3.3 Redundancy Removal via Subsumption Relations ... 107
7.4 The Painting Example ... 110
7.5 When Is the Constraint-Based Method Useful? ... 112
7.6 The COMBINE Algorithm ... 112
7.7 Related Work ... 115
7.7.1 Overview of Previous Work ... 115
7.7.2 Related Work by Smith and Peot ... 116
7.7.3 Related Work by Etzioni ... 118
7.8 Open Problems ... 118
7.9 Summary ... 119
7.10 Background ... 120

8. Plan Merging ... 121
8.1 The Value of Plan Merging ... 121
8.2 Formal Description ... 123
8.2.1 A Formal Definition for Plan Merging ... 123
8.2.2 An Example ... 125
8.2.3 Impact on Correctness ... 126
8.3 Complexity of Plan Merging ... 128
8.3.1 Deciding What to Merge ... 128
8.3.2 Deciding How to Merge ... 129
8.4 Optimal Plan Merging ... 130
8.5 Approximate Plan Merging ... 133
8.5.1 Algorithm APPROXMERGE ... 134
8.5.2 Example ... 134
8.5.3 Complexity ... 135
8.6 Related Work ... 136
8.6.1 Critics in NOAH, SIPE, and NONLIN ... 136
8.6.2 MACHINIST ... 137
8.6.3 Operator Overloading ... 138
8.7 Open Problems ... 138
8.7.1 Order-Dependent Cost Functions ... 138
8.7.2 Hierarchical Plan Merging ... 138
8.7.3 Enhancing Conflict Resolution ... 139

8.8 Summary ... 139
8.9 Background ... 139

9. Multiple-Goal Plan Selection ... 141
9.1 Consistency-Based Plan Selection ... 141
9.1.1 The Multiple-Goal Plan-Selection Problem ... 141
9.1.2 A CSP Representation ... 142
9.1.3 A Constraint-Based Solution ... 142
9.1.4 Evaluation—a Blocks-World Example ... 144
9.2 Optimization-Based Plan Selection ... 146
9.2.1 Examples ... 147
9.2.2 Complexity ... 148
9.2.3 A Heuristic Algorithm for Plan Selection ... 149
9.2.4 Examples ... 153
9.2.5 Empirical Results in a Manufacturing Planning Domain ... 155
9.3 Open Problems ... 156
9.4 Summary ... 157
9.5 Background ... 157

Part III. Hierarchical Abstraction

10. Hierarchical Planning ... 163
10.1 A Hierarchical Planner ... 164
10.1.1 Algorithm ... 164
10.1.2 Precondition-Elimination Abstraction ... 165
10.2 Specifying Refinement ... 171
10.2.1 Forward-Chaining, Total-Order Refinement ... 171
10.2.2 Backward-Chaining, Partial-Order Refinement ... 172
10.3 Properties of an Abstraction Hierarchy ... 174
10.3.1 Existence-Based Properties ... 174
10.3.2 Refinement-Based Properties ... 175
10.4 An Analytical Model ... 179
10.4.1 Assumptions ... 180
10.4.2 The Probability of Refinement ... 183
10.4.3 Analytical Result ... 184
10.5 Open Problems ... 187
10.6 Summary ... 187
10.7 Background ... 188

11. Generating Abstraction Hierarchies ... 189
11.1 Syntactic Connectivity Conditions ... 189
11.2 HIPOINT ... 192
11.2.1 ALPINE ... 194

11.2.2 Probability Estimates ... 196
11.2.3 Collapsing Nodes with Low Refinement Probabilities .. 199
11.2.4 Augmented Topological Sort of Abstraction Graph ... 200
11.3 Empirical Results in the Box Domain ... 200
11.4 Related Work on Abstraction Generation ... 203
11.5 Open Problems ... 204
11.6 Summary ... 206
11.7 Background ... 206

12. Properties of Task Reduction Hierarchies ... 207
12.1 Defining Operator Reduction ... 208
12.2 Upward Solution Property ... 213
12.2.1 Losing the Property ... 213
12.2.2 A Variable-Assignment Example ... 214
12.2.3 The Chinese Bear Example ... 214
12.3 Imposing Restrictions ... 216
12.3.1 Motivation ... 216
12.3.2 Unresolvable Conflicts ... 217
12.3.3 The Unique-Main-Subaction Restriction ... 217
12.4 Preprocessing ... 219
12.5 Modifying Operator-Reduction Schemata ... 219
12.6 Open Problems ... 221
12.7 Summary ... 223
12.8 Background ... 223

13. Effect Abstraction ... 225
13.1 A Motivating Example ... 225
13.2 Primary-Effect Restricted Planning ... 226
13.3 Incompleteness and Sub-optimal Solutions ... 229
13.4 An Inductive Learning Algorithm ... 230
13.5 How Many Training Examples ... 233
13.5.1 Informal Description ... 233
13.5.2 A Brief Introduction to PAC Learning ... 233
13.5.3 Application of the PAC Model ... 235
13.6 Open Problems ... 236
13.7 Summary ... 237
13.8 Background ... 237

References ... 239

Index ... 249

List of Figures

1.1 A painting example 2
1.2 A PERT network for a painting project 5
1.3 An outline of the issues addressed in this book 7

2.1 A partial-order plan 18
2.2 Two total orders of a partial order plan 19
2.3 Backward-chaining search in a space of plans 21
2.4 (a) Demotion, (b) Promotion 23
2.5 An initial painting plan 26
2.6 A one-step plan for painting 26
2.7 A two-step plan for painting 26
2.8 A final solution plan for painting 27
2.9 Forward search in a space of states 27
2.10 SatPlan solves a problem by verifying models on a timeline 36

3.1 A search tree explored by a planner 39
3.2 Search tree for a forward, total-order planner 40
3.3 Search tree for a backward, partial-order planner 43
3.4 Resolving a conflict by introducing a white knight step $\mathbf{s}_{wk}$ 45

4.1 A plan and its associated transitive closure matrix 52
4.2 An illustration of dynamic programming algorithms 54
4.3 An illustration of branch and bound algorithms 57
4.4 A CSP representation of the graph coloring problem 58
4.5 Revising the domain of a variable in a CSP 60
4.6 A search tree for a backtrack-based algorithm 62

5.1 Committing to a paint brush 68
5.2 The thrashing problem in search 69
5.3 A constraint satisfaction problem associated with a plan 71
5.4 A finite-domain plan where the variables form a constraint satisfaction problem 74
5.5 A final solution plan for the house-painting example 74
5.6 Comparing TopPop and FdPop 75
5.7 The variable-sized blocks domain 76

5.8 Performance in variable-sized blocks domain 77

6.1 Interactions in the household domain 88
6.2 Abstraction planning 96
6.3 Decomposition planning 97

7.1 Conflict 2 can be removed from a CSP by Var-Subsumption Theorem 108
7.2 A conflict-resolution constraint of Conflict 2 can be removed from the domain of Conflict 2 by Value-Subsumption Theorem 109
7.3 A CSP for the conflict-resolution in the painting domain 111
7.4 An operator graph for a simplified painting domain 116
7.5 An operator graph showing how a threat, from node 2 to node B, can be delayed 117

8.1 A variable-sized blocks world.................................. 122
8.2 Merging plan steps 124
8.3 A blocks world plan 125
8.4 Merged blocks world plan 126
8.5 An example where merging helps conflict resolution 127
8.6 Computational complexity of plan merging 129
8.7 Intuition of dynamic programming................................ 131

9.1 Comparison in blocks-world domain............................... 146
9.2 Search Space... 149
9.3 Illustrating S-connected classes................................. 152
9.4 A shopping example 154

10.1 Illustrating hierarchical planning 164
10.2 Two operators with the same abstract representation 166
10.3 A Towers of Hanoi domain....................................... 167
10.4 A simple travel domain... 170
10.5 Refinement of a total-order plan................................ 171
10.6 Upward-solution property 175
10.7 Our analytical model: a tree of abstract solutions, with a branching factor B and a depth $n+1$.......................... 179
10.8 Subproblems in a total-order refinement. 181
10.9 Subproblems in a partial-order refinement....................... 181

11.1 A plan segment and its refinement 191
11.2 Robot-box planning domain 194
11.3 Robot-box domain graph generated by ALPINE...................... 196
11.4 Robot box domain tests 202

12.1 An example operator reduction schema........................... 209
12.2 Restrictions on operator reduction schemas 209

12.3 Applying a reduction schema to a goal 210
12.4 A plan with unresolvable conflicts 214
12.5 (a) A variable-assignment plan containing unresolvable conflicts, (b) Resolving the conflict by reducing the plan in (a) 215
12.6 (a) A Chinese bear plan with unresolvable conflicts, (b) Resolving the conflicts by reducing the plan in (a) 216
12.7 Modifying a reduction scheme by removing some preconditions of an operator 221
12.8 Modifying a reduction schema by removing effects of operator α .. 222

13.1 A painting operator 225
13.2 Search trees of unrestricted planning (a) and primary-effect restricted planning (b) 228
13.3 Demonstrating problems of primary-effect restricted planning 230
13.4 Informal description of the learning process 234

List of Tables and Algorithms

2.1 Operator definition for the painting example 17
2.2 The POPLAN algorithm 20
2.3 Operator definition with resources 25
2.4 The TOPLAN algorithm 28
2.5 The SATPLAN algorithm 35

3.1 Summary of notations 42

4.1 The POPLAN algorithm 51
4.2 WARSHALL's algorithm 53
4.3 A depth-first search algorithm 53
4.4 The DYNAMIC algorithm 55
4.5 The BRANCH-AND-BOUND algorithm 57
4.6 The AC algorithm 61
4.7 The FWDCHECKING algorithm 63
4.8 The GSAT algorithm 65

5.1 Operator definition with resources 70
5.2 The FDPOP algorithm 73
5.3 Operator definition with resources 73
5.4 Operator definition with resources 76

6.1 The DECOMPLAN algorithm 87
6.2 The GDECOMP algorithm 93

7.1 Conflicts in the painting example 103
7.2 The COMBINE algorithm 113

8.1 The OPTIMALMERGE algorithm 133
8.2 The APPROXMERGE algorithm 135

9.1 The WATPLAN algorithm 143
9.2 The OPTIMALSELECT algorithm 150
9.3 Experimental results for Algorithm 4 using L_3 156

10.1 The HIPLAN algorithm 165

10.2 Operator definitions in the TOH domain 168
10.3 A criticality assignment in the TOH domain 168
10.4 Abstract operator definitions in TOH domain 169
10.5 Operators in a simple travel domain 170
10.6 The TotalRefine algorithm for precondition-elimination abstraction 173
10.7 The PartialRefine algorithm for precondition-elimination abstraction.. 173
10.8 Asymptotic search complexity for different regions when a solution exists 185

11.1 The Hipoint algorithm for creating a hierarchy 193
11.2 Operators for the robot-box domain 195
11.3 Robot-box domain hierarchy generated by Alpine............. 196
11.4 Algorithm Find-Probability 198
11.5 A matrix of refinement probabilities 199
11.6 Robot box domain hierarchy generated by Hipoint............. 200
11.7 CPU time comparison between Hipoint and Alpine in the box domain.. 202
11.8 An additional operator in the extended TOH domain 205

12.1 The HRefine algorithm 212
12.2 "fetch" example 219
12.3 The UmsCheck algorithm 220

13.1 Comparing unrestricted and primary-effect restricted planners ... 227
13.2 Operator definition, with selected primary effects, for the office domain example.. 228
13.3 The LearnPrim algorithm 232

1. Introduction

1.1 The Problem

Planning is about producing changes through actions. The planning tasks we face are, of course, diverse. What routes to follow to deliver an express package from one city to another? What actions to take in order to maximize one's investment rewards while reducing the amount of risk? What sequence of commands to give to a computer-controlled machine tool for manufacturing an automobile part? What update operations to follow for retrieving certain information from a large, distributed database? The list goes on.

With the advance of computer technology many planning activities can now be automated. We can envision a computer-aided planning toolkit for generating parts of a plan automatically, for selecting actions among a large set of alternatives, for looking for bugs in a complex plan, and for suggesting how to reduce the cost of a given plan.

The central aim of this book is to design a library of automatic planning methods using techniques in Artificial Intelligence. These methods are collections of planning algorithms resulting from extensive research in this area by the author and his co-workers, and are built on experiences that other Artificial Intelligence researchers have accumulated in the past several decades. All methods are aimed at making planning more *effective*. That is, the plan generation and reasoning process are to become more efficient, and the quality of the plans is to become higher (lower costs and risks, and higher reliability). Although these methods do not form a complete list of all algorithms needed for constructing an automatic planner, they nevertheless cover many of the important aspects of automatic planning. In the next section, we discuss these aspects at a fairly general level.

1.2 Key Issues

What problem-solving abilities must a planner possess in order to successfully compose a workable plan? We begin by considering the following scenario of a planning task: a robot is given the task of painting both a ceiling and a ladder. The robot formulates the following plan: it first fetches a can of paint

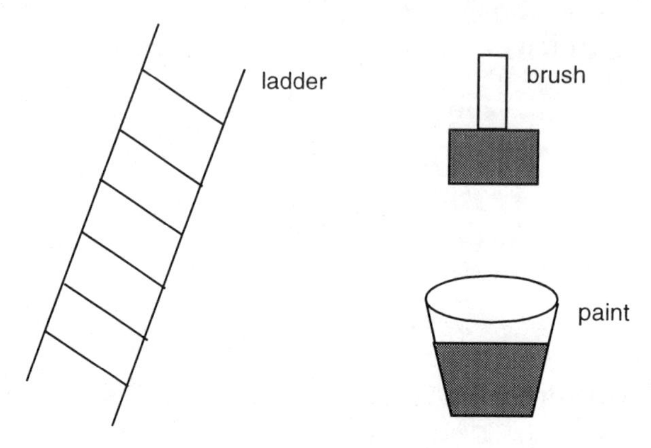

Fig. 1.1. A painting example

and two brushes. It then places the ladder under the ceiling and climbs up to the top. After painting the ceiling with one brush, it switches to another and proceeds to paint the ladder (see Figure 1.1). In this problem, we want the planner to generate a good plan, and to do so as quickly as possible.

The first issue we consider is efficiency. An intelligent planner must be able to compose a plan efficiently. The issue of efficiency is of major concern here because of the potentially large number of action combinations as plans; given ten actions that agents are able to perform, a naive planner might have to look through 10^{10} sequences of actions before finding a correct ten-step plan! An equally important issue is the quality of a plan generated, especially when the planner must choose among many alternative plans in order to find one with low cost. Here again the number of alternative plans might be prohibitive for a naive method to be applied. A planning tool that is able to compose a good plan quickly might be called an *effective planner*.

One way to construct a plan is incrementally; that is, add one plan step at a time until all final goals are achieved. Constraints can be posted in the process to ensure that the plan is free from harmful interactions among its steps. Finally, the plan steps could be added in either a forward or a backward manner. In the former case a plan grows from the current situation to the final situation, and in the latter the direction is reversed. These *incremental* methods to planning, on which virtually all AI planning research builds, might be called *basic planning algorithms.*

The central thesis of this book is that to be effective, a planning algorithm must go beyond a basic one. It should exploit the structures of the problem at hand and work along the lines of well-established human intuitions. One such intuition is *divide-and-conquer*, whereby a complex problem is divided into several more-or-less independent parts and those parts are solved separately. Another intuition is *hierarchical abstraction*, with which a distinction is made between parts of a problem that are difficult to solve and those that are relatively easy to solve. These parts are organized into the levels of a hierarchy, which can be used by a planning algorithm to make the planning process more effective. In this book we present formalized computational methods for realizing both of these intuitions.

Divide-and-Conquer. By the intuition of divide-and-conquer, an intelligent planner ought to have the ability to *decompose* a complex problem domain into subcomponents. *Problem decomposition* will make it possible to attack the individual problem components *separately*, thereby improving planning efficiency. In the painting example, the planner might decide that painting the ceiling and painting the ladder can be planned for separately. To do so incurs a number of intriguing issues. First of all, decomposition involves recognizing the parts of the problem domain that are independent to some degree. This process is nontrivial due to the potentially large number of possible decompositions; for a problem domain with ten possible descriptive features, there are potentially 2^{10}, or over one thousand possible ways just to divide these features into two groups! The second issue involves how the solutions to the individual components can be successfully combined. This again is a complex problem.

During the combination process, the need arises to reason about conflicts among multiple plans. For the painting robot, if the robot can only hold one brush at a time, then a resource conflict occurs between the part of the plan for painting the ceiling and the part of the plan for painting the ladder. Similarly, if the wet paint from painting the ladder precludes one from climbing up, then another conflict occurs because performing the former negates a precondition of the latter, which requires that the ladder be dry. With the number of conflicts mounting in a large plan, efficient methods for *conflict resolution* are necessary.

Another issue involved with the solution-combination process is that different solutions might contain redundant parts when combined. Some of these redundancies can be removed to further improve the overall quality of the plan. In the painting domain, if the two goals, painting the ceiling and painting the ladder, are planned for separately, the two plans, when combined, might contain two identical steps for fetching the ladder. These two steps can be merged into one. This leads to a strategy called *plan merging*, whereby parts of a plan can be merged together to yield a plan with lower cost.

Yet another issue in problem-decomposition is that a subproblem might have several alternative solutions. When the solutions to the subproblems are considered together again, the choice of one solution to a subproblem might affect how the solutions to other subproblems are selected. Some collective choices may lead to a globally incorrect plan; others may yield a dramatically more costly plan. This *plan selection* problem therefore requires an intelligent solution that is more than a random selection from each solution set.

Hierarchical Abstraction. As pervasive as divide-and-conquer, *hierarchical abstraction* is another often-used human intuition in attacking planning problems. This method essentially uses the distinctions between important aspects of a planning problem from those that are mere details. By so doing a planner could first work with a much smaller version of the problem and use the resultant solution to constrain subsequent problem solving. For our painting planner, an abstract plan might involve three actions at a high level of description in a sequence: get the paint and brushes; paint the ceiling; finally, paint the ladder. When all aspects of the problem are taken into account, these steps serve as a framework around which the entire plan could be fleshed out.

Improvements to the Basic Planners. Each of the above techniques, problem-decomposition and hierarchical abstraction, requires that a basic planner be available. In problem-decomposition, a basic planner is needed for solving each decomposed subproblem. In abstraction, a basic planner is needed for fleshing out a more abstract plan. It is therefore also an important task to improve the efficiency of the basic planners.

For a basic planner that works backwards from the goal, its efficiency is greatly affected by the number of actions that can achieve each given subgoal. The larger the number of such actions the more costly is the planning process. One way to reduce the number is to classify certain effects of an action as *primary effects.* These are the effects that determine the main purpose of an action. Others are considered as secondary. For example, "having the brush" is a primary effect of fetching the brush, and "the agent's hand is non-empty" might be considered secondary. By focusing on primary effects, a planner might lose its ability to generate plans for all problems, but we might still be happy as long as it could generate good plans for most of the problems more quickly.

Another technique we use to improve the efficiency of a basic planner is to consider available *resources* collectively. In the painting example, the planner might be facing a large number of available brushes for selection. If we do not have adequate knowledge about them initially, and if few of the brushes are useful in the end, then a naive commitment to a particular brush might result in a bad plan, a fact which may not be discovered until much later. An intelligent planner might decide to delay the choice on a particular brush, and choose one only when there is enough information for such a selection. This may allow a failed search path to be discovered early, and also allows many similar plans to be considered collectively.

In this book, we present these techniques in two separate chapters: Chapter 13 in Part I, and Chapter 5 in Part III.

1.3 Planning Versus Scheduling

A field closely related to planning is known as *scheduling.* Indeed, authors in Operations Research often use the two terms interchangeably. However, there is an important difference.

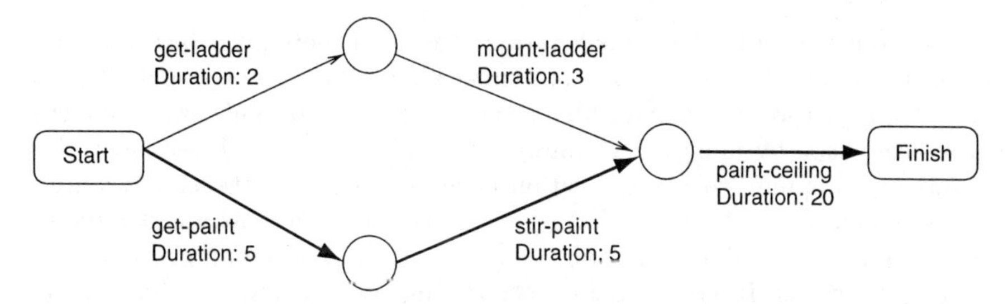

Fig. 1.2. A PERT network for a painting project. Each arc is an activity, and each node is a state (or event). The heavy lines illustrate a critical path

Scheduling grew out of efforts to formalize optimizations of large combinatorial problems. A characteristic scheduling problem is to assign machines to machined parts (e.g., automobile parts) given that each machined part could be handled by more than one machine. The complication arises from a number of factors: the difference in costs associated with each machine and machined part, various constraints on delivery deadlines and the availability of machines and storage resources. Often the aim of scheduling is to optimize certain value functions subject to a set of constraints.

One way to visualize the above optimization problem is to consider the computation being done on a network of tasks. Each task can be considered a high-level description of an action to be performed, yet without time and machine resources assigned to it. Although different assignment of resources can yield a solution with a different degree of satisfaction in the optimality criteria and constraints, the inter-relations between the tasks are more or less determined already. In other words, this network of tasks, often called a PERT (Project Evaluation and Review Technique) network, is *given* as input.

A PERT network is a set of tasks or activities constrained by a partial order. A *partial order* specifies which tasks must occur before which others. It is acyclic — if A is before B then B cannot be before A (see Figure 1.2). Many types of analysis can be performed on such a network; a typical one is CPM (Critical Path Method). In the figure, a critical path (boldfaced path) is determined by a longest path from the start state to the finish state. This path information is useful in determining the shortest time required to finish the project, should no unexpected events occur.

As emphasized above, finding a critical path in a PERT network assumes that a plan structure (a PERT network) is already given. Subsequent computation is performed to optimize a plan based on its structure. A natural question to ask, then, is how to construct the network of activities — a generalized plan — in the first place. This is the planning problem, a problem that has been more or less ignored by the field of Operations Research, but is nevertheless equally important and fascinating!

Obviously even for a plan generation algorithm there must be some form of initial input. We must know something about the general capabilities of the plan-execution agent, and about the outside world. This is exactly the approach taken by many AI planning researchers. As we will see later, this knowledge amounts to a specification of action representations and world-state specifications. When a plan is assembled, these actions might interact in different ways, calling for sophisticated methods for reasoning about the effects of actions. In contrast, in a scheduling problem, the action interactions are assumed known and sorted out; the only remaining task is to assign resources to each action to optimize a cost function. Reasoning about actions and their effects is one of the main distinctions between planning and scheduling.

It should also be noted that it may be very beneficial to interleave planning and scheduling in an intelligent way. A plan can be constructed partially when an assignment of resources on the planning tasks is attempted. This approach has the obvious advantage that if a plan cannot lead to a good schedule, it is abandoned early. It is nevertheless much clearer conceptually to draw a line between planning and scheduling; planning precedes scheduling. This is the approach taken in this book.

1.4 Contributions and Organization

This book introduces advanced computational methods for dealing with planning problems. We will strive for our methods to be as general as possible so that they can be applied under different underlying representational schemes. However, the real focus of our presentation will be on the algorithmic aspects of planning; we present intelligent algorithms for the above planning tasks and support our claim for their effectiveness theoretically and empirically.

In particular, the book is divided into the following three parts, illustrated in Figure 1.3.

Part 1: Basic Representations and Algorithms. In this part we present preliminary representations and basic algorithms for planning (Chapter 2). We also discuss some useful analytical methods for comparing planning methods (Chapter 3) and motivate our subsequent presentations. In addition, we give a systematic summary of useful supporting algorithms for plan reasoning (Chapter 4), so that the contents of the book are self-contained. Finally, we conduct a case study of an application of constraint satisfaction to planning, showing how natural it is to extend some basic planning algorithms to include more elaborate functionalities (Chapter 5).

Part 2: Planning Problem Decomposition. This part deals with the issue of problem decomposition and solution combination. The important questions are:

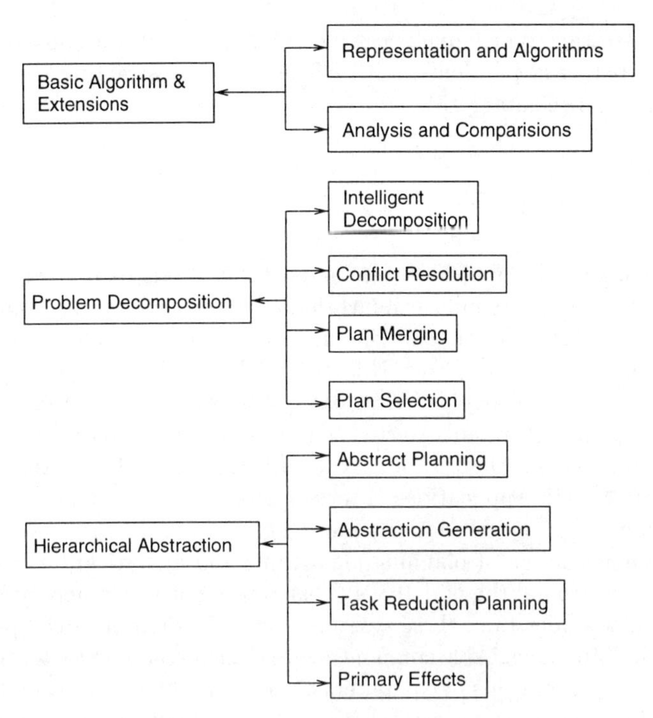

Fig. 1.3. An outline of the issues addressed in this book

- how to decompose a problem domain into separate parts, so that a plan can be made in each part concurrently (Chapter 6);
- how to resolve conflicts that arise among the different solution plans that are to be combined (Chapter 7);
- how to recognize opportunities for optimizing plans by merging repetitive actions (Chapter 8);
- how to select plans from a pool of potential candidates? The selection process may be guided by the need to resolve conflicts (Section 9.1), or by the need to optimize plans (Section 9.2).

Part 3: Planning with Hierarchical Abstraction. This part deals with a formalization of hierarchical planning and planning with abstraction. The important questions are:
- how to generate and refine plans in any given abstraction hierarchy (Chapter 10);
- how to automatically generate abstraction hierarchies by separating the important and unimportant aspects of a problem (Chapter 11);
- how to plan using a predefined library of task reduction schemata, and what properties of the library will ensure effective planning (Chapter 12);

– how to trade-off completeness (i.e., ability to find solutions for *all* solvable planning problems) with efficiency, by abstracting the effects of operators (Chapter 13).

1.5 Background

The problems of planning have fascinated generations of AI researchers. A main theme has been to build systems that could generate plans from scratch. Early work include Newell and Simon's GPS [101], Green's QA3 system for theorem proving [59], Fikes, Nilsson and Hart's STRIPS [46], Sussman's HACKER [130], and Warren's WARPLAN [143]. Around the 1970s, more sophisticated planning systems started to appear beyond research labs, including Sacerdoti's NOAH [114] Tate's NONLIN [132], Vere's DEVISER [140], and Wilkins' SIPE [147]. Applications of these systems ranged from office-building construction to space-craft activity planning.

The understanding of planning algorithms and analytical techniques began to mature around the mid-1980s. This movement has largely been driven by the need to understand these systems better, by dividing a complex planning system into manageable components, and then studying each component thoroughly. In other fields of computer science, one could, for example, spend an hour of classroom time in teaching *sorting algorithms* (see, for example, [2]) and then send the students off for implementation. But one could not do that with planning. The systems mentioned above required a certain degree of simplification and formal characterization first.

With this motivation, in his AI Journal paper in 1987, *Planning for Conjunctive Goals* [28], Chapman presented proofs that planning in general is undecidable, and that to decide whether a plan is correct is NP-hard (likely to require exponential time in the worst case). Pednault [103], working independently, provided a logically precise account of plan correctness under a general representation scheme, and discussed various ways of implementing search. Other treatments of this topic, to name a few, included work by Lansky [88], Drummond and Currie [39], Tenenberg [134], Minton, Bresina and Drummond [95], Gupta and Nau [60], McAllester and Rosenblitt [94], Bylander [24], Backstrom [14], Kambhampati [70], Knoblock [80], Barrett and Weld [17], and Yang [150] [1].

Today, several excellent surveys on planning are available. These include Wilkins' description of the SIPE system [147], Weld's article in AI Magazine [144], Allen, Hendler and Tate's collection *Readings in Planning* [5], book chapters in AI textbooks by Russell and Norvig [112], Dean, Allen, and Aloimonos [135], Winston [148], Rich and Knight [110], Tanimoto [131], and Charniak and McDermott [29].

[1] This list is by no means exhaustive; in subsequent chapters we will survey other relevant work in more detail.

By choosing to focus on algorithms and analysis techniques for plan reasoning, we could not manage to cover many other important topics in the scope of this book. These other topics include reactive planning, plan execution and monitoring [115, 1, 23], case-based planning [82, 163], planning in natural language understanding [145, 4], reasoning about actions and beliefs in logic [90, 61, 109], multi-agent planning [40], plan reuse and machine learning in planning [97, 73], GRAPHPLAN[21], and decision theory and control [35]. Each of these topics deserves thorough and complete coverage.

Part I

Representation, Basic Algorithms, and Analytical Techniques

Overview

The elements of strategy:
First, measurements
Second, estimates
Third, analysis
Fourth, balancing
Fifth, triumph

The situation gives rise to measurements
Measurements give rise to estimates
Estimates give rise to analysis
Analysis gives rise to balancing
Balance give rise to triumph

(Sun Tzu, The Art of Strategy, *Chapter Four)*

This part contains four chapters, serving to set the stage for the rest of the book. In Chapter 2 we first outline a simple language used to describe a planning domain, then describe some basic algorithms for constructing a plan. In Chapter 3 we show how to analyze and compare planning algorithms based on a few parameters. Throughout the book we use some traditional algorithms used in computer science, and for the book to be self-contained, these techniques are briefly reviewed in Chapter 4. Finally, in Chapter 5, we conduct a case study of the application of one of the supporting techniques, constraint satisfaction in planning. We show how to combine constraint satisfaction techniques with a basic planner, and illustrate the potential effectiveness of the resulting system.

2. Representation and Basic Algorithms

2.1 Basic Representation

One of the early representational methods for planning domains is known as the STRIPS representation [102]. This representation models actions as operations on a database, which records the current state of affairs in the world. Its simplicity is one of the major reasons for its vast popularity in many theoretical and practical work in planning. It has also been shown that many of the more exotic representational schemes, such as those involving a limited form of first-order logic representations, can be easily and naturally derived from the STRIPS representation. In this book, we will employ the STRIPS representation for the exposition of our major ideas and algorithms, and discuss possible extensions to these algorithms where the representational languages become more sophisticated.

2.1.1 Domain Description

A STRIPS domain description consists of a subset of first-order predicate language L for describing the domain, and an operator set $\mathcal{O}$ for describing the abilities of the agent. L is a restricted language consisting of predicate symbols p_i, negation $\neg$, constant symbols c_i, and variable symbols x_i. The *terms* of L are the constants and variables in L. An *atom* is an expression of the form

$$p(t_1, \ldots, t_n),$$

where P is an n-ary predicate and the parameters t_i are terms. The *ground* atoms are the atoms where all terms are constants. The *literals* (also called *propositions*) include all atoms and their negations. Further, for any literal p, $\neg\neg p$ is equivalent to p.

As an example, consider a description of a painting domain described in Chapter 1. Here the predicate symbols are

Painted, Havebrush, Havepaint, Haveladder, Dry, Handempty, ...

The constant symbols are

Door, Brush, Ladder, Ceiling, paint, ...

And the variables are ?brush, ?paint, ...

If one assigns truth values (TRUE or FALSE) to every literal in a domain description such that, whenever p is assigned TRUE, $\neg p$ is assigned FALSE, then one obtains a state. Because the number of literals needed for defining a single state is often too large to handle, we often talk about *state descriptions*. A state description is a subset S of all literals in a domain such that every literal in S is given a truth value "TRUE," their negations are false, and the rest of the literals in the domain definition are assumed unknown. In effect, S describes a set of states. With this machinery we can describe an initial state as a set of literals representing their conjunction. For example, in the painting domain an initial state description might be

$$\{\neg\text{Painted(Door)}, \neg\text{Painted(Ladder)}, \text{Dry(Ladder)}, \text{Handempty}\}$$

For succinctness, we often use the term *initial state* for an initial state description.

In STRIPS representation a *goal description* is also represented as a conjunction of literals. For example, in the painting domain, if we wish to have both the door and the ladder painted, we might say

$$\{\text{Painted(Door)}, \text{Painted(Ladder)}\}$$

where the comma "," represents logical and.

The second element, $\mathcal{O}$, in a STRIPS representation is a set of operators. Each operator α in O consists of four elements:

Operator Name. This is a list of symbols, OpName$(t_1, t_2, \ldots)$, where the first symbol is the name of the operator, and the rest are variables and constants appearing in descriptions of preconditions and effects of the operator.

Preconditions of α. This is a set of literals representing their conjunction, with the intention that these literals must be true immediately before the operator is applied. The preconditions of α are represented as Pre(α).

Effects of α. This is a set of literals representing their conjunction, with the intention that these literals will hold after the operator is applied. The effects of α are represented as Eff(α).

Operator Cost. This is a real value denoting the cost of performing the action denoted by the operator.

As an example in the painting domain, several operators are shown in Table 2.1.

If an operator description contains variables, it is called *partially instantiated.* If all variables are replaced by constants, then it is called a *ground instance* of the operator, and the instance of the operator is said to be *fully instantiated.*

Let S be a set of fully instantiated literals; S describes a set of states where all literals in S hold. Let s be a fully instantiated operator. s is *applicable*

Table 2.1. Operator definition for the painting example

Precondition	Effect	Cost
`get-brush`		
Handempty, Dry(?b)	Havebrush(?b), ¬Handempty	1.0
`paint-ceiling`		
Havebrush(?b), Dry(Ladder), Havepaint(`Paint`)	¬Dry(Ceiling), ¬Dry(?b), Painted(Ceiling)	2.0
`paint-ladder`		
Havebrush(?b), Havepaint(`Paint`)	¬Dry(Ladder), ¬Dry(?b), Painted(Ladder)	3.0
`return-brush`		
Havebrush(?b)	¬Havebrush(?b), Handempty	1.0

to S if all preconditions of s hold in S. The application of s to S results in another set T of literals describing the successor states of S reachable by s. T is defined as follows. Let Del be the set of literals l in S such that $\neg l$ is a member of Eff(s) (recall that $\neg\neg l = l$). That is, Del is the set of literals of S deleted by s. Then T is defined as:

$$T := (S - \text{Del}) \cup \text{Eff}(\mathbf{s})$$

T is called a successor state description of S after **s**.

A STRIPS planning problem is defined as a triple $\langle init, goal, \mathcal{O}\rangle$, where *init* is a set of initial state literals, *goal* are the goal literals, and $\mathcal{O}$ is the set of planning operators. A sequence of fully instantiated operators $\overline{\Pi}$ is called a *solution* to a STRIPS problem, where each operator is also known as a *step*. In this sequence, the first step is applicable to *init*, the ith step is applicable to the successor state description obtained after the $(i-1)^{th}$ step, and every literal in *goal* holds in the successor state description after the last step. $\overline{\Pi}$ is also called a total-order plan due to its sequential nature. A total-order plan satisfying the above definition is said to be *correct.*

2.1.2 Partial-Order and Partial-Instantiation Plans

Not every precedence ordering between plan steps in a total-order plan is necessary for maintaining its correctness. Sometimes a set of total-order plans can be compressed into a structure known as a partial-order plan. A partial-order plan consists of a *partially ordered set* of steps, along with a set of constraints on these steps. A step is a version of a planning operator in which some variables are replaced by constants. The steps are transformed from planning operators and inserted into the plan by a *plan generation*

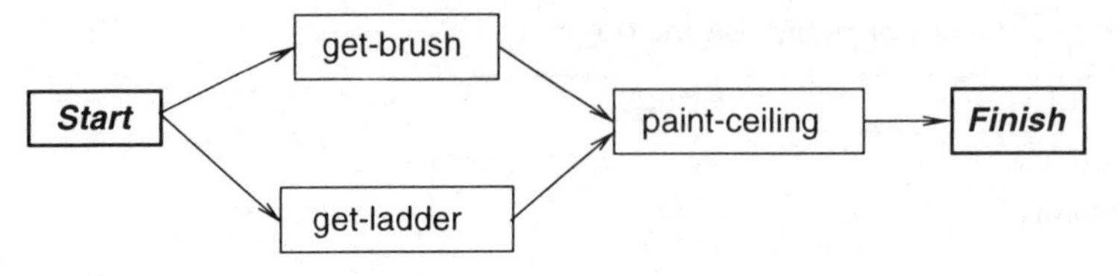

Fig. 2.1. A partial-order plan

algorithm. The ordering constraints state various precedence relation among the steps, and the variable-binding constraints state whether two variables, or a variable and a constant, can designate the same constant.

We describe each of these components in turn.

Step Orders. Mathematically, *a partial order* R on a set U is a binary relation such that

- *R is irreflexive.* That is, for every element a in U, (a, a) is not in R.
- *R is transitive.* That is, for all elements a, b, and c in U, if (a, b) and (b, c) are both in R, then (a, c) is also in R.

Graphically, a partial order R on U can be denoted as a network, where every element of U is a node and every pair in R is an edge. The definition of a partial order implies that the network does not contain any cycle.

In a partially ordered plan, the precedence relation on plan steps is a partial-order relation. As an example, Figure 2.1 is a partial-order plan for painting the ceiling. We use $\mathbf{s}_i \mapsto \mathbf{s}_j$ to denote that for steps $\mathbf{s}_i$ and $\mathbf{s}_j$, $(\mathbf{s}_i, \mathbf{s}_j)$ is a member of the partial-order relation. If $(\mathbf{s}_j, \mathbf{s}_i)$ is *not* a member of the partial-order relation, then we say that it is consistent for $\mathbf{s}_i$ to be before $\mathbf{s}_j$ in Π, and denote it by $\text{Consistent}_{\Pi}(\mathbf{s}_i \mapsto \mathbf{s}_j)$.

For simplicity, we also assume that in every plan there are two special steps $\mathbf{s}_{init}$ and $\mathbf{s}_{finish}$, where the former represents the initial state, and the latter the goal state. $\mathbf{s}_{init}$ is ordered before all other operators and has an empty set of preconditions. Its effects are the literals in the initial state. In a similar way, $\mathbf{s}_{finish}$ is ordered after all steps in the plan and has an empty effect set. Its preconditions are the goal literals.

A partial-order plan represents a set of total-order plans, where each total-order plan is a linear sequence of steps in the partial-order plan such that the partial-order relation in the latter is not violated by the sequence. Figure 2.2 shows two total order plans corresponding to the partial-order plan shown in Figure 2.1.

Variable-Binding Constraints. In addition to the ordering constraints between steps, a partial-order plan also contains a set of variable-binding constraints of the form $?x \neq ?y$, where $?x$ is a variable in some step in the plan and $?y$ is either a variable or a constant in the step. For example, in a painting plan a variable-binding constraint might be $?b \neq \text{Brush}_1$ where Brush_1 is a particular brush which might be unusable by the painting agent.

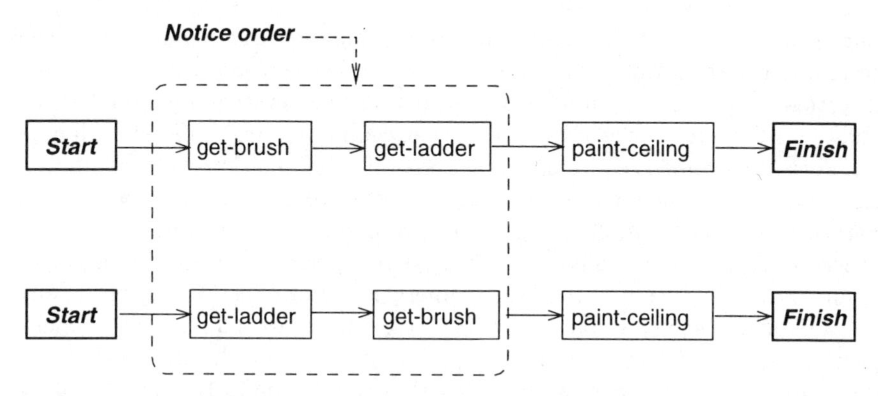

Fig. 2.2. Two total orders of a partial order plan

Another type of variable-binding constraint is of the form $?x = ?y$. These constraints force the instantiation of $?x$ to be the same as that for $?y$, and are known as *codesignation constraints.*

Causal Links. Finally, a partial-order plan might contain a set of *causal links*—relations between steps that serve as a record for precondition establishment. A causal link $\langle \mathbf{s}_i \xrightarrow{p} \mathbf{s}_j \rangle$ states that step $\mathbf{s}_i$ asserts a precondition literal p for $\mathbf{s}_j$. That is,

- p is a precondition of $\mathbf{s}_j$,
- p is an effect of $\mathbf{s}_i$, and
- $\mathbf{s}_i \mapsto \mathbf{s}_j$ holds in the partial order plan.

In a ceiling-painting plan, a causal link might be

$$\langle \texttt{get-brush} \xrightarrow{p} \texttt{paint-ceiling} \rangle,$$

where $p =$ Havebrush(Brush).

Putting it together, a partially ordered plan Π consists of a set of steps Steps(Π), partial ordering constraints Order(Π), variable-binding constraints Binding(Π) and a set of causal links C-Links(Π). If Order(Π) defines a partial-order relation on plan steps, then we say that the ordering relation in the plan is *consistent.* Likewise, if the variable-binding constraints Binding(Π) do not imply that, for some variable $?x$, $?x \neq ?x$, then we say that these constraints are *consistent.*

2.2 Basic Planning Algorithms

We will spend the rest of the chapter discussing plan generation algorithms. There are many ways to classify a planning algorithm, of which we will discuss four representative ways. These four classifications are made based on both the representation of a plan and the direction in which a partially completed plan grows. The first dimension of the classification distinguishes between total-order and partial-order plans. The second dimension distinguishes the direction of operator chaining; it is forward if new operators are added at the end of a plan, and backward if they are added at the beginning of a plan.

These two dimensions create four possible combinations. Depending on the choice, one can have a total-order, backward-chaining algorithm or a partial-order, backward-chaining algorithm. Likewise, one can have a a total-order, forward-chaining algorithm or a partial-order, forward-chaining algorithm. Each algorithm can have a different performance behavior in any given domain. Below, we only describe two of the four combinations, namely, partial-order, backward-chaining algorithms and total-order, forward-chaining algorithms. The other two combinations can be similarly constructed.

2.3 Partial-Order, Backward-Chaining Algorithms

One way to generate a plan for solving a STRIPS planning problem is to construct a partial-order plan by incrementally adding all plan components. In a basic implementation of a partial-order, backward-chaining planner, the preconditions of plan steps are achieved one at a time. In each iteration, a

Table 2.2. The POPLAN algorithm

Algorithm POPLAN($\mathcal{O}$, Π_{init});
Input: A set of planning operators $\mathcal{O}$, and an initial plan Π_{init} consisting of a start step and a finish step and a constraint that the start step be before the finish step;
Output: A correct plan if a solution can be found.

```
1   OpenList := { Πinit }.
2   repeat
3     Π:= lowest cost plan in OpenList;
4     remove Π from OpenList;
5     if Correct(Π)=TRUE then return(Π);
6     else
7       if Threats-Exist(Π) then
8         Let t be a threat; Succ := Resolve-Threat(t, Π);
9       else
10        Succ := Establish-Precondition(Π);
11      end if
12      Add all successors in Succ, generated in Steps 8 or 10, to OpenList;
13    end if
14  until OpenList is empty;
15  return(Fail);
```

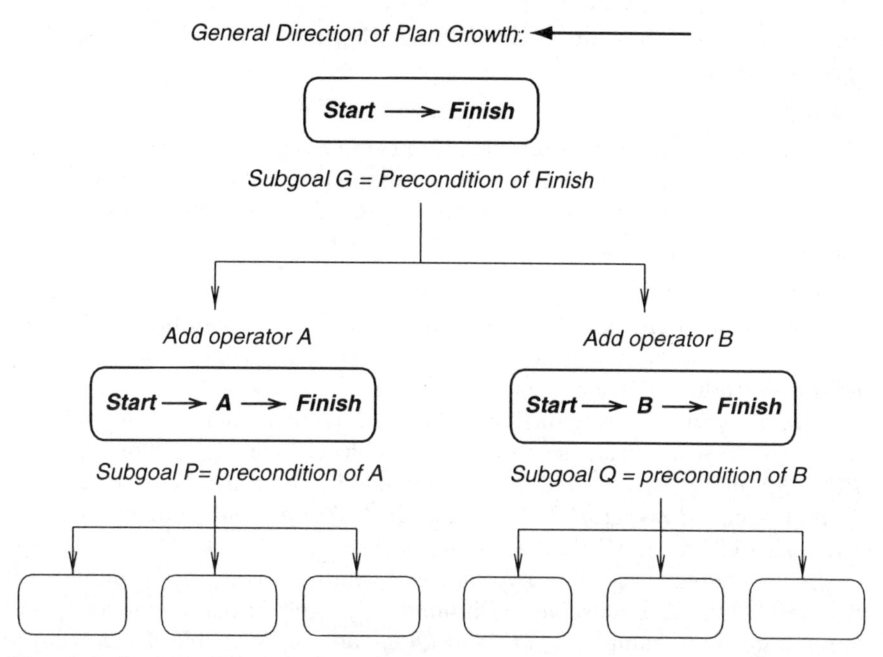

Fig. 2.3. Backward-chaining search in a space of plans

precondition of a step is selected to be achieved, and operators and plan steps that can achieve the precondition are identified. The precondition is then achieved by inserting a new causal link in the plan. Subsequently, threats to this causal link are removed. This process repeats until every precondition of every step in the plan has an associated causal link, and all negative threats (see below for a definition) in the plan are removed.

Table 2.2 shows a top-level version of a partial-order, backward-chaining algorithm. The initial plan consists of the start-step/finish-step pair. In steps 5, 7, 8 and 10, the algorithm uses a number of subroutines which we will flesh out in order of appearance. Figure 2.3 demonstrates the plan-generation process.

2.3.1 Correctness

The correctness of a partial-order plan can be defined in terms of that of a total-order plan. Recall that a partial-order plan Π corresponds to a set of totally ordered and fully instantiated plans. Let this set be Instances(Π). We say that a partially ordered plan Π is *correct* if the set Instances(Π) is nonempty, and if every member of Instances(Π) is correct.

The above definition can not be used directly to implement the subroutine `Correct()`, since in general the size of Instances(Π) might be very large; specifically, the number of instances is $n!$ if the partial-order plan has n

unordered steps. Thus, we resort to some other means of indirectly checking the correctness of a plan.

Open Preconditions.
One way to implement the subroutine `Correct()` is to rely on the information provided by causal links. First, we say that a precondition p of a step $\mathbf{s}$ in a plan Π is an **open precondition** if there does not exist a causal link $\langle \mathbf{s}_i \xrightarrow{p} \mathbf{s} \rangle$ in the plan.

Threats.
We now define a **threat** to a causal link as a step that can nullify the link. Threats come in two types, negative and positive. To define a threat, we first need to introduce the notion of **unification** of two literals.

Literals l_1 and l_2 are unifiable if they can be made identical by replacing some variables by either some other variables or constants. For example, $P(?x, ?y)$ and $P(\text{Brush}, \text{Ceiling})$ are unifiable by replacing $?x$ by Brush and $?y$ by Ceiling. However $P(?x, ?y)$ and $\neg P(?x, ?y)$ are not unifiable, nor are $P(\text{Brush}, \text{Ceiling})$ and $P(?x, ?x)$ be so.

In the context of a plan, two literals l_1 and l_2 might not be unifiable if the variable-binding constraints $\text{Binding}(\Pi)$ make two parameters of l_1 and l_2 separate. For example, $Q(?x)$ and $Q(?y)$ are not unifiable if $(?x \neq ?y) \in \text{Binding}(\Pi)$. If two literals p_1 and p_2 are unifiable by this definition, then we say that it is *consistent* that $p_1 = p_2$, or $\text{Consistent}_{\Pi}(p_1, p_2)$. Chapter 4 presents a more detailed overview of unification algorithms.

To define a negative threat, let $\langle \mathbf{s}_i \xrightarrow{p} \mathbf{s}_j \rangle$ be a causal link in a plan Π. Let $\mathbf{s}_k$ be a step in the same plan such that

1. it is consistent for $\mathbf{s}_k$ to be ordered between $\mathbf{s}_i$ and $\mathbf{s}_j$.
   That is, both $\text{Consistent}_{\Pi}(\mathbf{s}_i \mapsto \mathbf{s}_k)$ and $\text{Consistent}_{\Pi}(\mathbf{s}_k \mapsto \mathbf{s}_j)$ hold; and
2. there exists an effect q of $\mathbf{s}_k$ which can delete p.
   That is, $\exists q \in \text{Eff}(\mathbf{s}_k)$. $\text{Consistent}_{\Pi}(q = \neg p)$.

Then s_k is a threat to the causal link $cl = \langle \mathbf{s}_i \xrightarrow{p} \mathbf{s}_j \rangle$. The triple $[cl, s_k, q]$ is called a *conflict.*

Similarly, $\mathbf{s}_k$ poses a **positive threat** when its presence can make $\mathbf{s}_i$ useless. More precisely, let $\mathbf{s}_k$ be a step in the same plan such that

1. it is consistent for $\mathbf{s}_k$ to be ordered between $\mathbf{s}_i$ and $\mathbf{s}_j$.
2. $\mathbf{s}_k$ can reassert p.
   That is, $\exists r \in \text{Eff}(\mathbf{s}_k)$. $\text{Consistent}_{\Pi}(r = p)$.

One way to implement the subroutine `Correct()` is as follows:
`Correct(`Π`)` returns TRUE if the precedence relation and variable-binding relations are consistent, if there are no more open preconditions in the plan, and if there are no negative threats to any causal link in the plan. However, this is not the only way of defining `Correct()`. In Section 3.4.1 we will discuss another correctness-checking method.

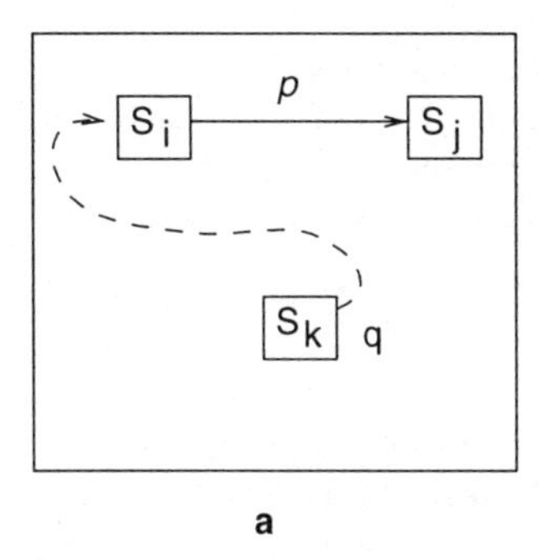

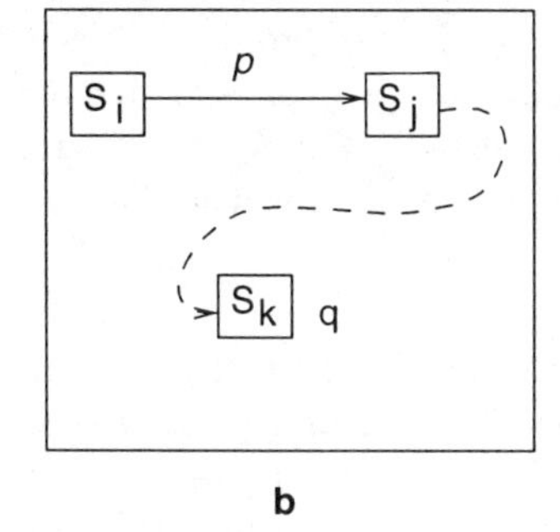

Fig. 2.4. (a) Demotion, (b) Promotion

2.3.2 Threat Detection and Resolution

Subroutine `Threats-exist()` can be implemented directly from the above definition of a threat. This subroutine takes a plan as input, and returns a set of (causal link, step) pairs where each step is a threat to the corresponding causal link.

To remove a threat from a plan Π, one can impose any one of the following constraints in the plan, as long as the resultant order and variable-binding relations are still consistent. Let $[\langle \mathbf{s}_i \xrightarrow{p} \mathbf{s}_j \rangle, \mathbf{s}_k, q]$ be a conflict.

Demotion of $\mathbf{s}_k$. This method corresponds to making $\mathbf{s}_k$ to be ordered before $\mathbf{s}_i$ (see Figure 2.4a). That is, add a constraint $\mathbf{s}_k \mapsto \mathbf{s}_i$ to the plan.

Promotion of $\mathbf{s}_k$. This method corresponds to forcing $\mathbf{s}_k$ to be ordered after $\mathbf{s}_j$ (see Figure 2.4b). That is, add a constraint $\mathbf{s}_j \mapsto \mathbf{s}_k$ to the plan.

Separation. This method introduces a variable-binding constraint so that $\neg p$ and q will never be the same. Suppose that for some i, x_i is the ith parameter of p and y_i the ith parameter of q. Also suppose that either x_i or y_i is a variable. Separation corresponds to adding a variable-binding constraint $x_i \neq y_i$. This will make it impossible for p and q to clash due to a threat.

Each of these threat-resolution methods can introduce one or more successor plans. Subroutine `Resolve-Threat`(t, Π) is implemented so that it returns the set of all successor plans as a result of resolving a threat t.

2.3.3 Establishing Preconditions

To achieve an open precondition p of a step $\mathbf{s}$, we first scan the steps in Π for possible candidate steps for achieving p. Let $\mathbf{s}_e$ be a step in Π with an effect q such that

1. $\text{Consistent}_{\Pi}(\mathbf{s}_e \mapsto \mathbf{s})$, and
2. $\text{Consistent}_{\Pi}(q = p)$.

Then $\mathbf{s}_e$ is a candidate *establisher* or *producer* of p for $\mathbf{s}$.

Likewise, we also search the set of planning operators in $\mathcal{O}$ to find candidate establishers for p. Let α be an operator with an effect q, such that p and q are unifiable. A unifier for p and q is a set of pairs

$$\theta = \{\langle ?x_1, t_1 \rangle, \langle ?x_2, t_2 \rangle, \ldots, \langle ?x_n, t_n \rangle\}$$

where each $?x_i$ is a variable, and each t_i is either a variable or a constant. *Applying* θ to p and q means that every occurrence of $?x_i$ in p or q is replaced by the corresponding term t_i. After the *substitution* operation, p and q look identical. Once the unifier is found, we can apply it to the entire operator schema and plan structure to obtain an instance of the operator and plan, respectively. For a more detailed introduction to unification, see Chapter 4.

Based on unification, the subroutine `Establish-Precondition` is implemented as follows. For each candidate establisher $\mathbf{s}_e$, we generate a successor plan as an instance of Π. The set of all such instances constitutes the successor plans for Π. More specifically, let θ be a unifier of p and q. A successor is obtained as follows:

1. apply θ to Π and $\mathbf{s}_e$,
2. if $\mathbf{s}_e$ is a new step, add $\mathbf{s}_e$ to Steps(Π),
3. add ordering constraints $\mathbf{s}_e \mapsto \mathbf{s}$ and $\mathbf{s}_{init} \mapsto \mathbf{s}_e$ to Order(Π),
4. let p' be an instance of p obtained by applying θ to p. Add $\langle \mathbf{s}_e \xrightarrow{p'} \mathbf{s} \rangle$ to C-Links(Π).

Subroutine `Establish-Preconditions` returns the set of all successor plans generated in this manner.

Properties of Planning Algorithms.
Two formal properties have been used most often in judging planning algorithms. A planner satisfies the **Soundness Property** if every solution Π that is output by the planner is *correct.* An orthogonal property is the *Completeness Property*, requiring that if a solution exists for a planning problem $\langle$ *init*, *goal* $\rangle$, then the planner will terminate with a solution (although not necessary the optimal one).

Suppose that, in addition to completeness, we further require that, for every planning problem $\langle$ *init*, *goal* $\rangle$ that has a solution, a planner can find the least-cost solution. Then the planner is called **admissible**.

To show that a planner is both sound and complete, a useful proof method is *induction*. For our partial-order planner POPLAN, if a least-cost-first search strategy is used in selecting a plan from the open list (Step 3 in Table 2.2, assuming all operators have positive costs), then by induction it can be shown to be both sound and complete.

Soundness and completeness, however, are often guaranteed with a heavy toll in search efficiency. Completeness, for example, sometimes requires that the planner plow through the entire search space without missing a single frontier. The over-cautiousness is likely to cause exponential growth in the size of the search frontier, or the open-list. Several alternative methods have been suggested to get around this problem. One is to settle for approximate solutions, solutions that are sometimes incorrect (see, for example [57]) or highly likely to be correct [156]. Another is the infusion of strong domain knowledge for pruning portions of the search space that do not contain solutions. A third method is to use local-search based planning algorithms, which we will discuss at the end of the chapter.

2.3.4 A House-Painting Example

We now illustrate the POPLAN algorithm with a painting example, where our goals are to paint a ceiling, a door and a table. To achieve these goals, suppose also that we have available the operators as shown in Table 2.3. In the initial state, we state that all brushes are dry:

$$\text{Dry(B1)}, \text{Dry(B2)}, \text{Dry(B3)}$$

We wish to achieve the goals

$$\text{Painted(Ceiling)}, \text{Painted(Door)}, \text{Painted(Table)}$$

To begin with, a plan with a start step and a finish step is generated (see Figure 2.5). Suppose that goals are achieved in a top-down order. Then Painted(Ceiling) is the first subgoal selected, and the operator `paint-ceiling` is converted into a step to be inserted into the plan (see Figure 2.6). Next an open precondition Dry(?cbr), of `paint-ceiling`, is selected. To achieve this subgoal a new causal link is created from the start step to support `paint-ceiling`.

When the next subgoal Painted(Door) is achieved, a `paint-door` step is inserted into the plan, as shown in Figure 2.7. This step becomes a threat to the causal link $\langle \mathbf{s}_{init} \xrightarrow{p} \texttt{paint-ceiling} \rangle$, where $p = \text{Dry(B1)}$. To resolve

Table 2.3. Operator definition with resources

`paint-ceiling`(?cbr, Ceiling)		
Preconditions	Effects	Cost
Dry(?cbr)	Painted(Ceiling), ¬Dry(?cbr)	5.0
`paint-door`(?dbr, Door)		
Preconditions	Effects	Cost
Dry(?dbr)	Painted(Door), ¬Dry(?dbr)	5.0
`paint-table`(?tbr, Table)		
Preconditions	Effects	Cost
Dry(?tbr)	Painted(Table), ¬Dry(?tbr)	5.0

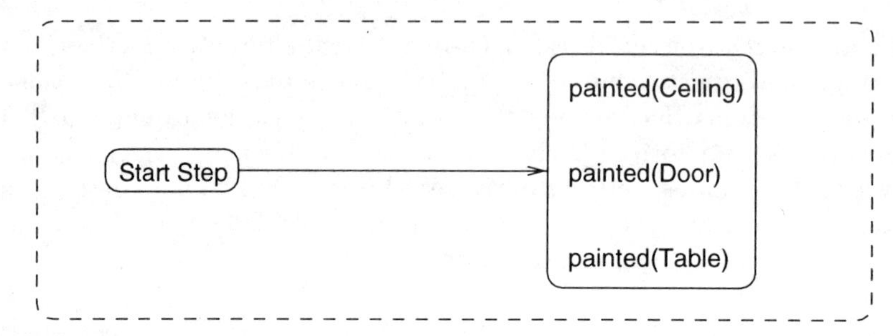

Fig. 2.5. An initial painting plan

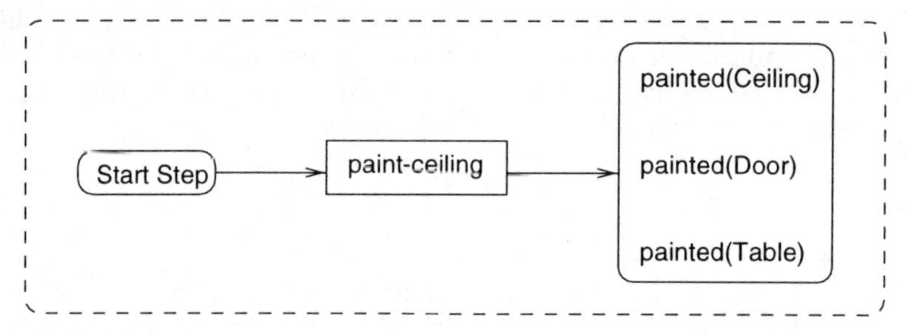

Fig. 2.6. A one-step plan for painting. The variable *?cbr* is instantiated to B1

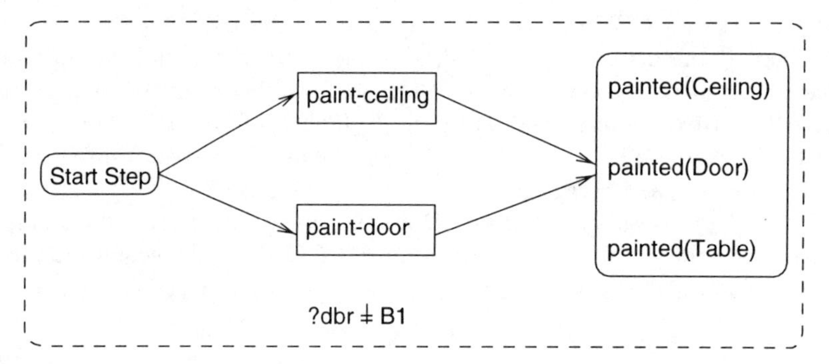

Fig. 2.7. A two-step plan for painting. A separation constraint is added to resolve a conflict

this conflict, a separation constraint is suggested by the planner, creating an inequality B1 $\neq$?dbr.

When the last goal Painted(Table) is achieved, and when the two threats in the plan are resolved, the plan is as shown in Figure 2.8. In the final plan, every precondition has an associated causal link, and no threats remain. This plan chooses paint brush B1 to paint the ceiling, B2 to paint the door, and B3 to paint the table. Because each operation relies on a different paint brush, they can be executed in parallel.

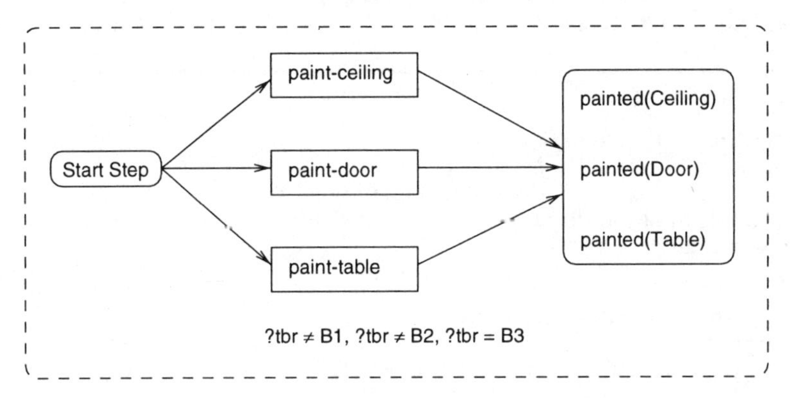

Fig. 2.8. A final solution plan for painting

2.4 Total-Order, Forward-Chaining Algorithms

A forward-chaining algorithm starts from an initial state description and searches for a goal in a space of domain situations (see Figure 2.9). Each situation is a description of the world in which an agent might find itself. Fully instantiated steps are added one at a time to the end of a plan, and a total-ordered solution plan is returned whenever a goal is reached.

In the implementation of forward-chaining shown in Table 2.4, a **node** encodes the information along a search path. Specifically, a node N_{fc} is a tuple:

$$N_{\text{fc}} = \langle \text{state, parent node, } \mathbf{s}, \text{ cost} \rangle$$

In this representation, the state is a current state description of the agent, the parent node is a node which leads to the present node by one step, **s**. The cost is the total cost of the path up to the current node, possibly augmented with some heuristic information. For each node N_{fc}, State-of(N_{fc}) returns the state element of that node.

In step 3 of the ToPlan algorithm a node (and the corresponding state) is selected to be *expanded*, resulting in a set of successor nodes to be gener-

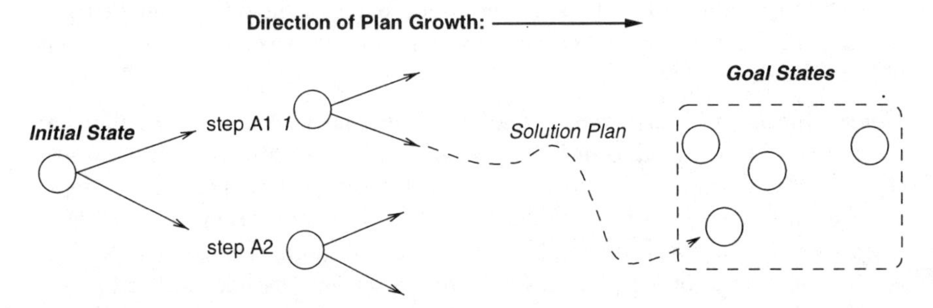

Fig. 2.9. Forward search in a space of states

Table 2.4. The TOPLAN algorithm

Algorithm TOPLAN(O,$State_i$,$State_g$);
Input: A set of planning operators O,
an initial state $State_i$ and a goal state description $State_g$
Output: A totally ordered, fully instantiated plan if a solution can be found. Otherwise, Fail.

```
1    OpenList := {⟨State_i, ∅, ∅, 0⟩}.
2    repeat
3      N_fc := lowest cost node in OpenList;
4      remove N_fc from OpenList;
5      if State_g ⊆ State-of(N_fc) then return (Solution-Path(N_fc));
6      else
7        generate successor nodes of N_fc from the applicable operators in O;
8        for each successor node Succ do
9          if State-of(Succ) is not previously expanded then
10           add Succ to OpenList;
11         else
12           compare the cost of Succ to previous path;
13           if Succ has lower cost then
14             add Succ to OpenList;
15           end if (step 13)
16         end if (step 9)
17       end for
18     end if (step 5)
19   until OpenList is empty;
20   return(Fail);
```

---

ated. To expand a node, all operators in $\mathcal{O}$ are searched to find those whose preconditions match the state of N_{fc}. Each instance is applied to yield a successor node. When a goal state is encountered, a function Solution-Path is applied to trace the plan steps from the goal back to the root node, returning the inverse of the step sequence. This sequence is a total-order solution plan.

As with partial-order plans, by induction one can show that the TOPLAN algorithm is both sound and complete when plans are selected from the open-list in a least-cost-first order.

Conversion from Total-Order to Partial-Order Plans. A total-order plan can be converted to a partial-order plan by removing constraints that are not necessary for maintaining its correctness. Backstrom [15] presents an in-depth analysis of this conversion problem as well as of several other previous algorithms for solving this problem. He shows that obtaining the least-constrained version of a total-order plan is NP-hard. However, one can still employ a greedy algorithm to find a sub-optimal version of the partial-order plan from a total-order solution. For example, one can incrementally take away the ordering constraints between pairs of plan steps. A step-order $\mathbf{s}_i \mapsto \mathbf{s}_j$ can be removed if after the removal, the plan remains correct. One can repeat this procedure until no more ordering constraints can be further removed without making the plan incorrect. The plan thus obtained is a partial-order plan, although it may not be the least-constrained partial-order plan.

2.5 More Advanced Operator Representations

The above definitions and algorithms are all based on the basic STRIPS representation. Extensions to more excessive representations turned out to be quite natural and straightforward. In this section we consider one extension in planning operator representation, and discuss how the basic plan-generation algorithms can be extended accordingly. In the rest of the book, we always base our discussion on the STRIPS representation, mainly for its simplicity. The extension presented here is based on the UCPOP representation and algorithm which are designed and implemented by Barrett and Weld [105].

The STRIPS representation is sometimes called *propositional*, since it permits only conjunctive literals as operator preconditions and effects. In many practical situations, however, the propositional representation is not sufficient. One often needs to consider universally quantified preconditions and effects, as well as conditional effects.

Consider the house painting example. Suppose that one requirement for painting the ceiling is that all furniture in the room is covered with drop-cloth. Instead of stating Covered(F) for each individual piece of furniture F, we can use a variable $?f$ to denote any piece of furniture, and state

$$\forall\ ?f\ s.t.\ \text{Inroom}(?f, \text{Room}).\ \text{Covered}(?f) \tag{2.1}$$

In English, this sentence is read "for all $?f$ such that Inroom($?f$, Room) is true, Covered($?f$) holds". If in a room there are many pieces of furniture, this representation could save a lot of effort in domain encoding. Universal quantification could be similarly extended to operator effects.

We could also allow operator effects to be *conditional*, in the form of

eff **when** *cond*

This means that when *cond* is true, *eff* will also be true after the application of the operator. As an example involving both generalizations of effects, suppose that in the painting domain, we state a fact that after painting, all the rooms that are painted in a building B would smell of paint. We could say

$$\forall\ ?r\ s.t.\ \text{Inbuilding}(?r, B).\ (\text{Smells}(?r)\ \textbf{when}\ \ \text{Painted}(?r)) \tag{2.2}$$

The meaning of this first-order sentence is that for all rooms $?r$ in a building B, $?r$ smells after it is painted.

Assume that all predicates hold for a finite, static set of objects. The basic planning algorithms are quite straightforward to extend in order to account for the universally quantified operator representations [105]. Suppose that each universally quantified variable x satisfies some predicate T, such that $T(x)$ holds for a finite, static set of objects. T is the predicate Inroom in equation (2.1), and the predicate Inbuilding in equation (2.2). The objects o_i are assumed to be finite in number, and are declared in the initial state as $T(o_1), T(o_2), \ldots, T(o_n)$. In equation (2.1) they are different pieces of furniture and in equation (2.2) they are the individual rooms in a building. Then for each universally quantified precondition of the form

$$\forall ?x \ s.t. \ T(?x). \ P(?x)$$

we explicitly state all $P(o_i)$ literals as a conjunction:

$$P(o_1), P(o_2), \ldots, P(o_n)$$

As such the POPLAN algorithm need not change much in order to accommodate for universal quantification.

For conditional effects, the backward-chaining partial-order planning algorithm can be extended as follows. During planning, procedure `Threats-Exist` looks for threats to a causal link $cl = \langle \mathbf{s}_i \xrightarrow{p} \mathbf{s}_j \rangle$. A step $\mathbf{s}_k$ is a (*negative*) threat to cl if $\mathbf{s}_k$ could possibly be between $\mathbf{s}_i$ and $\mathbf{s}_j$, and

(a) an effect literal of $\mathbf{s}_k$ possibly denies p, **or**
(b) $\mathbf{s}_k$ has a conditional effect ($\neg$*eff* **when** *cond*), where *eff* and p unify.

Case (a) above could be handled by the original algorithm, while case (b) serves as part of the extension.

For case (b), a method known as *confrontation* could be used. Specifically, the conditional effect states that $\neg$*eff* holds when *cond* holds. Therefore one way to remove the threat is to make *cond* false. This could be ensured by adding $\neg$*cond* as an open precondition of the step $\mathbf{s}_k$.

Likewise, for `Establish-Preconditions` to handle conditional effects, we make the following change. To achieve a precondition p, the procedure searches for an operator or a step with an effect e or a conditional effect of the form (e **when** c), where e unifies with p. In the latter case, c will be added as a precondition of the step that establishes the new causal link for p.

To sum up, the foregoing discussion demonstrates that the partial-order planning algorithm requires only minimal change when the domain language is extended to more expressive forms. The same claim holds for total-order planners.

2.6 Search Control

A plan-generation process can be considered as a search in a space of plans. The root node of the search space corresponds to an initial plan. At any step of the search process, a set of successor nodes is generated to expand the current search horizon. Search is successful if a node is found that contains a correct plan.

For both total-order and partial-order planning, the search space can be quite large. The problem of controlling the search is therefore very important. In general, the search-control problem occurs in two dimensions. In the first, a decision has to be made as to which node among the set of all frontier nodes should be selected next for expansion. A naive strategy would be to order the frontier nodes by the total cost of operators in a current plan; a plan with the minimal cost is selected next.

A second dimension of search control is defined as the problem of selecting a next "flaw" to work on, given that a frontier node or a plan has been selected. Here a flaw is either an open precondition or a threat to be resolved. Extensive research in planning has shown that the order in which the flaws are resolved often has a profound effect on search time and space [107].

Following the development of UCPOP, Peot and Smith [107] suggested the following strategy:

1. if there exists a threat that has zero or one resolution option, it is selected;
2. else, if an open precondition exists, achieve the open precondition;
3. otherwise, all threats have more than one resolution option. These threats are called *unforced.* Arbitrarily select an unenforced threat to resolve.

This strategy is called "delay unforced threat," or DUnf.

A problem of the DUnf strategy is that in the last step, no preference is specified as to which unenforced threat to resolve next. This problem is fixed by Joslin and Pollack [68] in a generalized strategy called LCFR (least-cost flaw repair). The repair cost of an open precondition or a threat is defined to be the number of nodes generated as a result of repairing it. LCFR will select a threat with minimal repair cost. Experiments show that in many domains LCFR is indeed an effective heuristic to use [68]. The LCFR strategy is later extended by Srinivasan and Howe [127].

Attempting to maintain focus on the achievement of a particular open condition or the resolution of a threat, Schubert and Gerevini [116] proposed a zero-commitment, last-in first-out selection strategy (ZLIFO). Given a plan, the strategy works as follows:

1. if there is a threat in the plan, select the threat to work on (however, delay separation). If there is more than one threat to select from, select the one most recently added in the plan (LIFO);
2. if there is a unachievable or uniquely achievable open condition in the plan, select the condition;

3. otherwise, select an open precondition that is most recently added to the plan (LIFO).

The strategy derives its name from the fact that Step 2 above corresponds to a zero-commitment choice, and that the selection method is based on the well-known LIFO strategy. The ZLIFO strategy was found to be very effective in a number of test domains.

2.7 Satisfaction Based Methods

2.7.1 Overview

The above methods are called systematic planning algorithms because they add actions in a plan incrementally and explore the space of plans systematically. In contrast, Kautz and Selman have developed a local-search based method for plan generation. This method is a randomized procedure for solving satisfiability type of problems in combinatorics. It is shown to outperform the systematic planners in several domains.

Generally speaking, a systematic search method starts from an empty plan, and repeatedly adds actions and constraints to a partially completed plan. The process continues until all goals are achieved and the entire plan becomes correct. A local-search algorithm, however, often starts from a randomly generated solution. The solution might contain some conflicts or errors due to unachieved preconditions or missing causal-links. A local-search process will then remove these errors and conflicts. In each step of the search process, a part of the solution is "repaired" so as to minimize the number of errors in a new solution.

Central to the local search approach to planning is the ability to form a solution, although the solution can be incorrect. However, before planning is completely done, there is usually no solution for one to work with; among other things, we do not know how many steps will be required in the final solution plan. Kautz and Selman solved the problem by using a clever idea, derived from Blum and Furst's GRAPHPLAN [21]. They consider the world as progressing through a sequence of time points $t = 0, 1, 2, \ldots, k$. Each time point gives rise to a state, which is a completely specified and instantiated set of propositions. Then, assuming that the plan would take n steps for a fixed integer n, an attempt is made to verify that a correct plan of n steps can be found, leading from the initial state to the state at time point n. The plan should achieve the goal formula no later than the last time point n. If no such plan is found within a certain upper bound on the amount of computation, n is incremented and the local-search process repeated.

To implement the above idea, three key elements are needed. First, a method for specifying what a correct solution plan is must be designed. This specification is also known as a *domain model*, describing the consistency constraints and actions for a state. The domain model is indexed on time point t. Second, a method must be designed for first instantiating the domain model at every time point up to and including time point n, and then converting the domain model to a uniform representation that is acceptable by a local search algorithm. A local search algorithm (for example, GSAT; see Chapter 4) is then activated. This step is also known as *model satisfaction.* If successful, model satisfaction produces a sequence of states. Finally, the sequence of states are converted to a sequence of actions. This action sequence corresponds to a solution plan.

2.7.2 Model Definition

To motivate the satisfaction-based methods, consider an example of moving a book from someone's home to a university. We can describe this domain using the following predicates. Connected(`loca`, `locb`) means that locations `loca` and `locb` are connected. We do not include a time index for Connected because it does not change in different states. An object can be moved between any two connected locations. At(`object`, `loc`, i) means that at time point i, object `object` is located at location `loc`. Inside(`object`, `container`, i) means that at time point i, `object` is inside `container`.

Following a state-based encoding of actions, we first describe the facts that hold in each state at every time point (thus all formulas below are universally quantified).

– *no object can be at two different locations at the same time.*

$$(\text{At}(\texttt{object}, \texttt{loca}, i) \wedge (\texttt{loca} \neq \texttt{locb})) \Rightarrow \neg \text{At}(\texttt{object}, \texttt{locb}, i)$$

– *no object can be at a location and inside a container at the same time.*

$$\neg(\text{At}(\texttt{object}, \texttt{loca}, i) \wedge \text{Inside}(\texttt{object}, \texttt{container}, i))$$

– Home *and* University *are always connected.*

$$\text{Connected}(\text{Home}, \text{University})$$

– *no object can be inside two different containers at the same time.*

$$\neg(\text{Inside}(\texttt{object}, \texttt{container}_a, i) \wedge \text{Inside}(\texttt{object}, \texttt{container}_b, i) \wedge (\texttt{container}_a \neq \texttt{container}_b))$$

Next, we describe all ways in which a literal "flows" from one time point to the next, and we do this for all literals in the domain:

– *if an object is at a location, it either remains at that location or goes into a container that is also at that location.* The latter possibility describes the action "load object into a container."

$$\begin{array}{l}\text{At}(\texttt{object},\texttt{loc},i)\Rightarrow\\ \quad \text{At}(\texttt{object},\texttt{loc},i+1)\vee\\ \quad \exists\ \texttt{container} \in \{\text{SchoolBag},\text{BriefCase}\}.\\ \qquad (\text{At}(\texttt{container},\texttt{loc},i)\wedge\\ \qquad \text{Inside}(\texttt{object},\texttt{container},i+1))\end{array}$$

– *if a container is at a location, it either remains at that location or is transported to a connected location.* The latter possibility describes the action "move container to another location."

$$\begin{array}{l}\text{At}(\texttt{container},\texttt{loc},i)\Rightarrow\\ \quad \text{At}(\texttt{container},\texttt{loc},i+1)\vee\\ \quad \exists\texttt{locb} \in \{\text{Home},\text{University}\}.\\ \qquad (\text{Connected}(\texttt{loc},\texttt{locb})\wedge\\ \qquad \text{At}(\texttt{container},\texttt{locb},i+1))\end{array}$$

– *if an object is in a container, it either remains in that container or is located at the same location as the container.* The latter possibility describes the action "unload object from container."

$$\begin{array}{l}\text{Inside}(\texttt{object},\texttt{container},i)\Rightarrow\\ \quad \text{Inside}(\texttt{object},\texttt{container},i+1)\vee\\ \quad \exists\texttt{loc} \in \{\text{Home},\text{University}\}.\\ \qquad (\text{At}(\texttt{container},\texttt{loc},i)\wedge\\ \qquad \text{At}(\texttt{object},\texttt{loc},i+1))\end{array}$$

The above model of the domain constrains the truth values of various propositions. For example, if an object is placed inside a container, the second state axiom makes sure that the object is no longer at its original location. This constraint ensures that the object follows the container wherever the latter moves.

The initial state includes all facts true at time point 0. For the above example, these facts are: At(SchoolBag, Home, 0), At(BriefCase, Home, 0), At(Book, Home, 0) and Connected(Home, University).

Similarly, the goal is a logical formula which states what is expected to be true at the last time point. It can also be a statement of a constraint on the behavior of the agent; for example, a goal might state that a certain formula must be true at the second to last time point. The last time point is of course unknown before the problem is solved. So we can initially set the point to 0 and then increment it repeatedly. For the book-carrying problem above, we could set it to be at the last time point n:

$$goal = \text{At}(\text{Book},\text{University},n).$$

We can then systematically push the envelope defined by n from 0 onward, and verify the logical model at each intermediate time point.

2.7.3 Model Satisfaction and Plan Construction

To verify that a model exists up to and including time point i using local search algorithms, we must convert the logical model to a form acceptable by the search algorithm.

First, each universally quantified variable such as `loc` and time points i are instantiated to constants. For example, a formula of the form $\forall\, i.\ \mathrm{At}(A, B, i)$ is instantiated into

$$\mathrm{At}(A, B, 0) \wedge \mathrm{At}(A, B, 1) \wedge \mathrm{At}(A, B, 2) \wedge \ldots \mathrm{At}(A, B, n)$$

The above axioms can be converted to a logical form known as *conjunctive normal form*; the conversion is straightforward (see [102], p. 145 for an algorithm). After this conversion we obtain a set of clauses, each clause being a set of literals that are fully instantiated. The set of clauses represents the conjunction of the clauses in the set, and each clause represents a disjunction of the literals in it.

Let the set of clauses representing the entire model at time point i be $\mathrm{CSet}(i)$. Let the set of all clause sets up to and including time point n be $\Sigma(n)$:

$$\Sigma(n) = \{\mathrm{CSet}(0) \cup \mathrm{CSet}(1) \cup \ldots \mathrm{CSet}(n)\}$$

We can then invoke the SATPLAN algorithm described in Table 2.5. In this algorithm, the input parameter MaxEnv represents the maximum number of plan steps we are willing to examine.

Table 2.5. The SATPLAN algorithm

Algorithm SATPLAN();
Input: A set clauses $\mathrm{CSet}(i)$, a parameter MaxEnv .
Output: a sequence of states if successful; Fail otherwise.

```
1    for n := 0 to MaxEnv do
2       decide the maximum number of tries and flips;
3       call GSAT(Σ(n)) (GSAT is described in Chapter 4);
4       if GSAT returns a satisfying assignment T then
5          convert T to a sequence of states S;
6          determine actions that go between the states (see below);
           let resulting plan be Π;
7          return Π;
8       end if
9    end for
10   return(Fail);
```

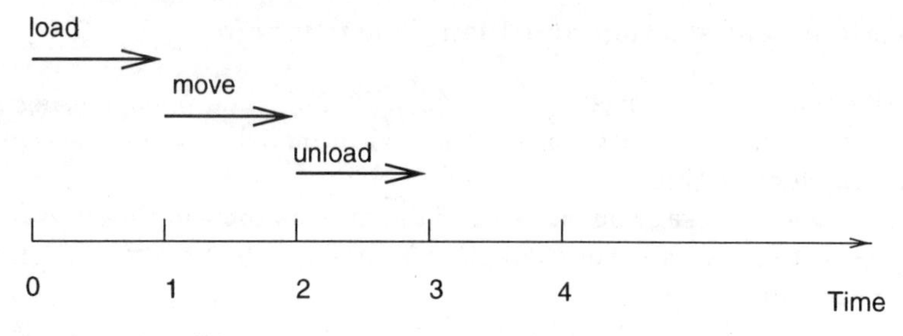

Fig. 2.10. SatPlan solves a problem by verifying models on a timeline

There are two possible outcomes for the algorithm: either the model is satisfiable within a specified time bound, or it is not. In the former case, we are provided with a sequence of states, starting from the time point 0 and ending at e, where e is the last time point for the particular model. The next task is to construct an action sequence from the state sequence.

For the book-carrying example the time points in a satisfied model are shown in Figure 2.10. Corresponding to each time point is a state:

$$\texttt{state}_0 = \left\{ \begin{array}{ll} \text{At(Book, Home, 0),} & \text{At(SchoolBag, Home, 0)} \\ \text{At(BriefCase, Home, 0),} & \text{Connected(Home, University)} \end{array} \right\}$$

$$\texttt{state}_1 = \left\{ \begin{array}{ll} \text{Inside(Book, SchoolBag, 1),} & \text{At(SchoolBag, Home, 1)} \\ \text{At(BriefCase, Home, 1),} & \text{Connected(Home, University)} \end{array} \right\}$$

$$\texttt{state}_2 = \left\{ \begin{array}{ll} \text{Inside(Book, SchoolBag, 2),} & \text{At(SchoolBag, University, 2)} \\ \text{At(BriefCase, Home, 2),} & \text{Connected(Home, University)} \end{array} \right\}$$

$$\texttt{state}_3 = \left\{ \begin{array}{ll} \text{At(Book, University, 3),} & \text{At(SchoolBag, University, 3)} \\ \text{At(BriefCase, Home, 3),} & \text{Connected(Home, University)} \end{array} \right\}$$

From the above sequence of states, we could obtain a set of adjacent states

$$(\texttt{state}_0, \texttt{state}_1), (\texttt{state}_1, \texttt{state}_2), (\texttt{state}_2, \texttt{state}_3).$$

Each pair can then be used to select an action responsible for the transition. This can be done by applying actions, one at a time, to the initiating state, and comparing the resulting state with the second element in each pair. If the resulting state corresponds to the second element, the action is chosen. To do this, however, requires that we have a representation of the individual actions. For the book-carrying domain, our actions are listed below.

`load` :

$$\begin{array}{l} \text{At}(\texttt{object}, \texttt{loc}, i) \wedge \\ \text{At}(\texttt{container}, \texttt{loc}, i) \Rightarrow \\ \quad \text{Inside}(\texttt{object}, \texttt{container}, i+1) \end{array}$$

unload :

$$\text{Inside}(\texttt{object}, \texttt{container}, i) \wedge$$
$$\text{At}(\texttt{container}, \texttt{loc}, i) \Rightarrow$$
$$\text{At}(\texttt{object}, \texttt{loc}, i+1)$$

move :

$$\text{At}(\texttt{container}, \texttt{loca}, i) \wedge$$
$$\text{Connected}(\texttt{loca}, \texttt{locb}) \rightarrow$$
$$\text{At}(\texttt{container}, \texttt{locb}, i+1)$$

After the computation is done, we end up with a plan:

$$\texttt{load}(\text{Book}, \text{SchoolBag}, 0)$$
$$\downarrow$$
$$\texttt{move}(\text{SchoolBag}, \text{Home}, \text{University}, 1)$$
$$\downarrow$$
$$\texttt{unload}(\text{Book}, \text{SchoolBag}, 2)$$

Kautz and Selman report that for large-scale blocks-world problems and a logistics planning problem, SATPLAN is several orders of magnitude faster than planning algorithms that are based on systematic search methods [77]. In the same set of experiments, several other variations of GSAT are applied, including the Davis–Putnam procedure developed by Crawford and Auton [33] and the Walksat algorithm [120].

2.8 Background

The domain and operator language in this chapter are adapted from Chapman's TWEAK [28], which uses a version of STRIPS domain language [46]. The POPLAN algorithm with its partial-order plan representation is based on the SNLP algorithm [94, 17]. The TOPLAN algorithm is a simplified version of the A* algorithm described in most AI textbooks [102].

How to implement a correctness-checking algorithm for a partial-order plan has been a subject of ongoing investigation. Chapman defines a modal-truth criterion for TWEAK [28]. The criterion was considerably simplified in an implementation of ABTWEAK, a hierarchical version of TWEAK [159, 149, 160] . The SNLP algorithm [94, 17] introduces another level of simplification in Chapman's criterion, by eliminating the so-called *white knights* altogether (see Chapter 7). A thorough discussion can be found in [75]. Kambhampati presents an extension to a backward-chaining, partial-order planner, using multi-contributor causal structures [71]. Veloso and Blythe discuss the need for combining total-order and partial-order planning in a single framework [139]. Also see [129] for a discussion for the need for different heuristics to guide a planner. Joslin and Pollack distinguish between active and passive commitments in constraint posting, and propose as well as evaluate a technique for representing and reasoning about all planning decisions with constraints [69].

The more elaborate operator language in Section 2.5 is based on the UCPOP formalism [105], which is a partial-order implementation of Pednault's ADL language for planning domain description [104]. For total-order planners, the extension can be done similarly. See, for example, [10, 19] for a description of the TLPLAN planner, which encodes domain knowledge for controlling a total-order, forward-chaining planner.

Finally, there has been a new thrust in planning with non-STRIPS style action encoding. Examples are GRAPHPLAN [21] and local-search based methods [77]. SATPLAN, described in the last section, is adapted from an AAAI 96 paper by Kautz and Selman [77]. Due to space limitations, we only include one form of state-based encodings in which no explicit propositions for actions are used. Other forms of encodings that include actions as a form of propositions have been attempted. It should be noted that so far no one approach yet appears best for all domains.

3. Analytical Techniques

3.1 Basic Analytical Techniques

Plan generation can be viewed as a search in a space of nodes. This is true for both forward-chaining and backward chaining planners, and for total-order and partial-order planners.

Consider the two planners that we discussed in the previous chapter. For a forward-chaining planner, a node records state information, and the arcs connecting the states are formed by operators applicable to the originating states. For a backward-chaining planner a node is a plan, and the arcs connecting the nodes are plan-modification operations applicable to a plan. In both cases, a planner explores a *tree* of nodes. This is known as the search tree of a planner.

In a search tree, the number of nodes, together with the amount of processing time on an individual node, give rise to an estimation of the time complexity. We number the level of nodes in a search tree starting from the root node; the root has a level number zero. The children nodes of the root are at level one, and so on. For each node at level i of the search tree, let B_i be the maximum number of arcs originating from the node. Let D be the minimum number of arcs explored by the planner on a path leading from the root (see Figure 3.1). D could be used to measure the search-tree depth under

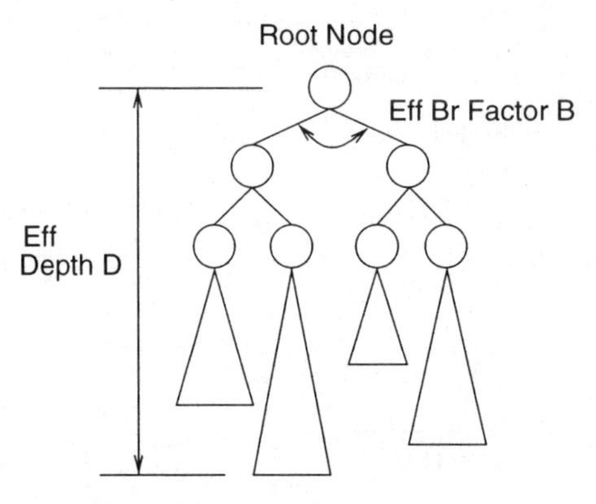

Fig. 3.1. A search tree explored by a planner

the assumption that the planner will not wander beyond this depth during a search process. Also let T_{node} be the maximum amount of time spent by the planner on each node. Then the worst-case time complexity of a total-order or partial-order planner could be bounded by

$$\text{Time}(\texttt{PLANNER}) = \sum_{i=0}^{D}((\prod_{j=0}^{i} B_i) * T_{node}) \tag{3.1}$$

To estimate the time complexity of a specific planner, we must flesh out three factors, B_i, D and T_{node}.

3.2 Analyzing Forward-Chaining, Total-Order Planners

The branching factor of TOPLAN is determined by the number of applicable operator *instances* to each state. The effective branching factor B_f is set to be the maximum such number. Notice that the number of applicable operator instances could be much larger than the number of planning operator templates themselves, since each planning operator template could be instantiated to many instances. As an example, consider planning a path in a building from one room to another. The domain might have only one operator template for moving. But for each room, the number of instances is determined by the number of adjacent rooms.

The search depth of TOPLAN is determined by the minimum number of plan steps leading from the root node to an optimal goal, which we denote by N. Figure 3.2 shows the parameters.

The time spent on each node is determined by the time needed to unify an operator's preconditions to the literals in a state, and by the time spent

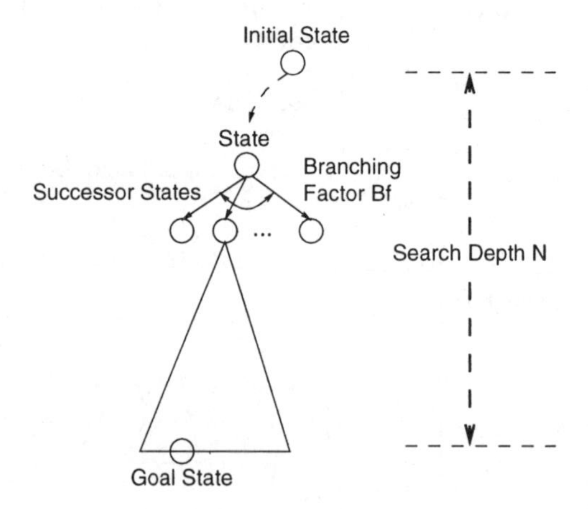

Fig. 3.2. Search tree for a forward, total-order planner

to generate successor states and checking whether these states have been generated before. This time, T_f, is assumed constant throughout the entire search space.

Putting it together, we obtain a worst-case time-complexity formula for forward, total-order search:

$$\text{Time}(\text{ToPlan}) = O(B_f^{\text{N}} * T_f) \tag{3.2}$$

The formula points out one potential shortcoming of forward, total-order search. The branching factor is determined by the number of applicable operators to each state, and this number could be extremely large. To see this, consider building a wooden table from a pile of lumber. Each applicable tool that could change the state of wood, ranging from drills and saws to hammers and nails, can potentially give rise to a large number of applicable operators. The branching factor in this case could even be infinite.

It is due to shortcomings like this that early AI researchers have turned away from pure forward-chaining planners. An alternative is backward-chaining planners. In these planners, the branching factor is determined by the number of ways to achieve a goal. This number could be potentially much smaller than the number of operators applicable to a state; in general, the number of ways to make a hole in a piece of wood is smaller than the number of ways to shape the same piece of wood. We will analyze backward planners in more detail in the next section.

3.3 Analyzing Partial-Order Planners

Recall that in the POPLAN algorithm, described in Chapter 2, the following components need be specified:

step 5. `Correct(Π)`, as a termination condition;
steps 7 & 8. `Threats-Exist` and `Resolve-Threat`, for detecting and resolving threats in a plan;
step 10. `Establish-Precondition`, for building a causal link for an open precondition.

In the POPLAN algorithm, B is the maximum number of successor plans generated either after step 8, or after step 10. The depth D is the maximum number of plan expansions in the search tree from the initial plan state to the solution plan state. Let P denote the maximum number of preconditions or effects for a single step, and let N denote the total number of plan steps in an optimal solution plan (except the start and finish steps).

To flesh out the effective branching factor B, we first define the following additional parameters. We use B_{new} for the number of new operators found by step 10 for achieving p, B_{old} for the number of existing plan steps found by step 10 for achieving p, and r for the number of alternative constraints for removing one threat. The effective branching factor of search by the POPLAN algorithm is either $B_{new} + B_{old}$, or r since each time the main routine is followed, either step 8 is executed for removing threats, or step 10 is executed to build causal links. If step 8 is executed, r successor states are generated. Otherwise, $(B_{new} + B_{old})$ successor plan states are generated.

Next, we expand the effective depth D. In a solution plan, there are $N * P$ number of (p, s_{need}) pairs, where p is a precondition for step s_{need}. Let f be the fraction of the $N * P$ pairs chosen by step 10. Let v be the total number of times any fixed pair (p, s_{need}) is chosen by step 10. Finally, for each pair (p, s_{need}) to be achieved, a set of threats is accumulated to be removed. Let t be the number of threats generated by step 10. A summary of the parameters can be found in Table 3.1.

The total time-complexity formula can now be obtained. Since the number of nodes in a tree is roughly the number of leaf nodes, we count the number of leaves in the search tree (see Figure 3.3). Step 10 for achieving open preconditions is executed $f * N * P * v$ times, each time generating $(B_{new} + B_{old})$ successor plan states. For each of these states, step 8 for resolving threats is executed t times, each time generating an additional r successor plans. Putting these together in equation (3.1), we get

$$\begin{aligned} \#nodes(\text{POPLAN}) &= \sum_{i=0}^{f*N*P*v} \prod_{j=0}^{i} (B_{new} + B_{old}) * r^t \\ &= O\left((B_{new} + B_{old})^{f*N*P*v} * r^{f*N*P*v*t}\right) \\ &= O\left([(B_{new} + B_{old}) * r^t]^{f*v*N*P}\right) \end{aligned} \quad (3.3)$$

To obtain the time complexity, we must determine the time complexity for evaluating the termination condition, `Correct`(Π). For illustrative purposes, we consider the following version of the termination condition:

1. every precondition of every step must have a causal link,
2. no negative threats exist for any causal link.

Table 3.1. Summary of notations

B	effective *branching* factor
D	effective search *depth*
T_{node}	average time per node
N	total number of operators in a plan
P	total number of *preconditions* per operator
f	*fraction* of (p, s_{need}) pairs examined
v	average number of times a (p, s_{need}) pair is *visited*
t	average number of *threats* found at each node
r	average number of ways to *resolve* a threat
B_{new}	average number of *new* establishers for a precondition
B_{old}	average number of existing (or *old*) establishers for a precondition

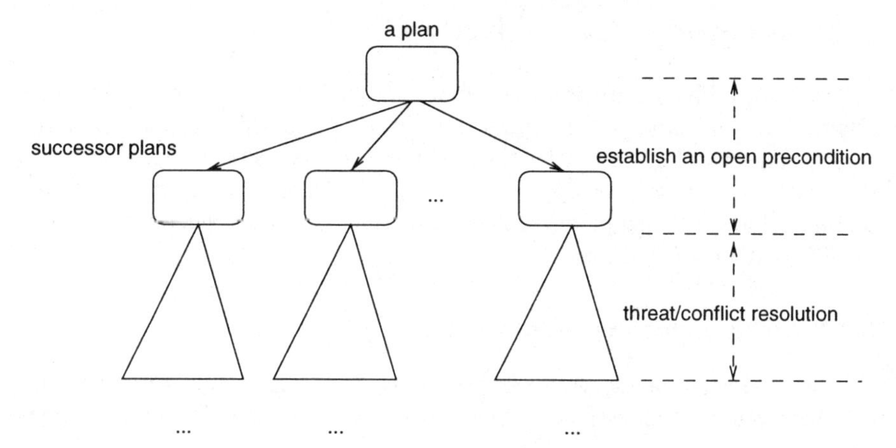

Fig. 3.3. Search tree for a backward, partial-order planner

To check this condition efficiently, one can maintain a list of threats and open preconditions. When this list is empty, the termination condition is true. Thus, the time per node is constant, or $T_{node} = O(1)$.

Also, to establish causal links, every precondition in a solution plan must be visited once. Thus, $f = 1$. In addition, since every precondition p of every step s_{need} has an associated causal link, (p, s_{need}) needs not be visited more than once. Therefore, $v = 1$. For this version of planner, the time complexity for checking termination condition is $N * P$. Let the branching factor be B_b:

$$B_b = [(B_{new} + B_{old}) * r^t]^P \tag{3.4}$$

And the time complexity for POPLAN is

$$\text{Time}(\text{POPLAN}) = O\left({B_b}^N\right) \tag{3.5}$$

We now consider a comparison of this formula with the time complexity for a forward-chaining total-order planner (formula (3.2)). In both planners, the effective search depths are the same. To make the comparison simple, we also assume that the time-per-node factor T_f is a constant. A total-order planner will take more time to find a solution if its branching factor B_f is larger than B_b for a backward-chaining, partial-order planner. This is the situation in a domain where the number of ways to modify a state is large (hence B_f is large), and the B_b factor is small. The latter is true when the number of ways to achieve a goal is small (hence B_{new} and B_{old} are small), and the number of ways to resolve the threats in a plan is small (hence r is small). On the other hand, a forward-chaining total-order planner will take less time than a backward-chaining, partial-order one when the number of operators orginating from a state is small, and the number of ways to achieve a goal and to resolve threats in a plan are large.

3.4 Case Study: An Analysis of TWEAK

Instead of building explicit causal links for each precondition, Chapman's TWEAK[28] uses what is known as the Modal Truth Criterion to verify the correctness of a plan. In this section we describe a simplified implementation of TWEAK, to which we apply the analytical formula derived in the last section. This simplified planner is the basis for an abstract version of TWEAK known as ABTWEAK[160].

3.4.1 An Implementation of TWEAK

We focus on an implementation of the modal truth criterion. Consider a precondition p for a plan step s_{need}. p is called *correct* if the following conditions are true:

Establisher. There exists a plan step s_{add} such that s_{add} is an establisher of p for s_{need}. To qualify as an establisher, s_{add} must be ordered before s_{need} in Π, and must have an effect p.

Conflict-Free. All clobbering steps are taken care of.
A step $\mathbf{s}_k$ qualifies as a clobbering step if it is not ordered after s_{need}, and one of its effects q could unify with $\neg p$. By this definition, all negative threats are clobbering steps.

To take care of a clobbering step $\mathbf{s}_k$ for the pair (p, s_{need}), one of the following must be true. Let s_{add} be an establisher of p for s_{need}.

Demotion of $\mathbf{s}_k$. $\mathbf{s}_k$ is ordered after s_{need}. That is, a constraint $s_{\text{need}} \mapsto \mathbf{s}_k$ exists in the plan.

Promotion of $\mathbf{s}_k$. $\mathbf{s}_k$ is ordered before s_{add}. That is, $\mathbf{s}_k \mapsto s_{\text{add}}$ in the plan.

Separation. Suppose that for some i, $?x_i$ is the ith parameter of p and $?y_i$ the ith parameter of q. Also suppose that either $?x_i$ or $?y_i$ is a variable. Separation corresponds to a variable binding constraint $?x_i \neq ?y_i$. This will make it impossible for p and q to clash, by denying each other.

White Knight. Let $\mathbf{s}_{wk}$ be a plan step such that it is consistent that $\mathbf{s}_{wk} \mapsto s_{\text{need}}$ and $\mathbf{s}_k \mapsto \mathbf{s}_{wk}$. Furthermore, $\mathbf{s}_{wk}$ has an effect e such that e and p unify. A white knight step is one which provides a shield between the step s_{need} and the clobbering step $\mathbf{s}_k$; whenever p is in danger, $\mathbf{s}_{wk}$ comes to its rescue.
More precisely, the *white knight* constraints are
- $\mathbf{s}_{wk} \mapsto s_{\text{need}}$,
- $\mathbf{s}_k \mapsto \mathbf{s}_{wk}$, and
- either p or $\neg q$ is an effect of $\mathbf{s}_{wk}$.

The first three constraints are familiar to us by now. The fourth one, the white knight constraint, deserves a little more explanation.

White knight is a concept formalized by Chapman. It is plausible that a plan step $\mathbf{s}$ in a plan might be able to achieve a precondition of another

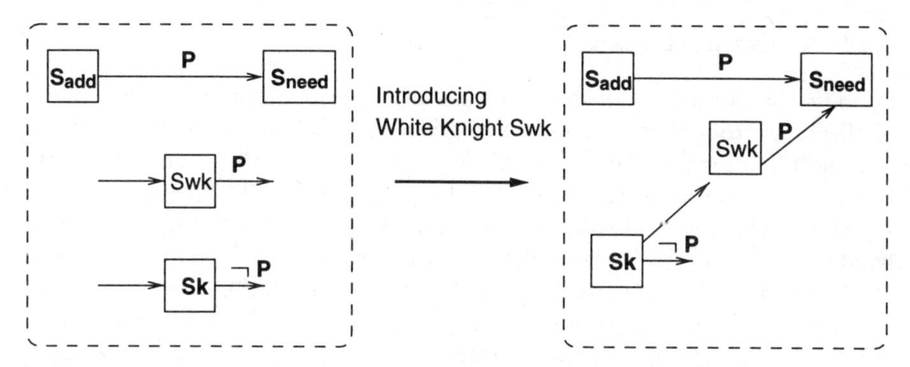

Fig. 3.4. Resolving a conflict by introducing a white knight step $\mathbf{s}_{wk}$

step in the plan. In such cases, and when a threat is discovered, it is possible to impose further ordering and variable-binding constraints onto the plans, such that the step **s** serves as a *shield* for the threatened precondition, against the clobberer. This is the process of introducing white knights, shown in Figure 3.4.

It should be pointed out that it is necessary to include white knights for conflict resolution in a planner. A partial-order causal-link planner such as SNLP which does not use white knights is known to be complete; it can find a solution to a problem if there is one. A white knight simply provides a combination of promotion, demotion, separation, and precondition establishment methods used in SNLP [94, 17]. Although it is not necessary for ensuring completeness, it does provide a heuristic advantage in some circumstances. One reason why it is used in TWEAK is because it provides a kind of short cut in a search tree; instead of imposing constraints one at a time in each search node, a group of constraints are imposed all at once. It is also true that including the white knight in search increases the branching factor of search by one. Therefore, whether a planning method using white knights will be more efficient than one not using it depends on the balance between reducing the path lengths along some search-tree branches and increasing the branching factor.

Let our implementation of the modal truth criterion be referred to as MTC. Given a plan, MTC(*plan*) returns TRUE if all preconditions of all plan steps in Π are correct. If so then TWEAK can terminate (Step 5 in the POPLAN algorithm, Table 2.2). Otherwise, some plan step s_{need} must exist for which the above conditions are false. In that case, MTC(Π) can return a (p, s_{need}) pair. This is the subgoal to be established in step 10 of the POPLAN algorithm. The rest of the algorithm could remain the same.

3.4.2 Analyzing TWEAK

Suppose that for each plan step, the number of preconditions and the number of effects are roughly the same. For a given plan Π, the MTC checks every precondition of every plan step. This would take $O(N * P)$ in the worst case. For each precondition, the white knight part of the MTC requires that a search be made for clobberers, and for each clobber, another search for white knights. Thus, for each precondition, checking the MTC conditions takes $O(N * P)^3$ time. This is the T_{node} factor in equation (3.1).

Since the same precondition might be clobbered more than once, v(TWEAK), the average number of times a (p, s_{need}) pair is visited, is greater than one. Also, not every precondition will be selected to achieve; if p is already true by virtue of the initial state, for example, then it might be the case that p will be left alone. Therefore, $f(\text{TWEAK}) < 1$.

Putting this together, the time complexity for TWEAK is as follows: Let B_b be $[B_{new} + B_{old}\ ) * r^t]^P$.

$$\text{Time(TWEAK)} = O\left({B_b}^{N*f*v} * (N * P)^3\right) \tag{3.6}$$

Judging from the analytical formulas for the causal-link planner POPLAN and the non-causal-link TWEAK, we see that neither is absolutely better than the other. From equation (3.6) it appears that TWEAK spends more time on each plan. However, because of the f and v factors, the search depth of TWEAK may turn out to be less than N. This happens when $f \ll 1$ and $v = 1$, a situation where many of the goals and preconditions need not be explicitly established. In contrast, POPLAN has a smaller processing time on each plan. Thus, in other domains POPLAN can be better than TWEAK.

3.5 Critics of Basic Planners

Comparing the worst-case behavior of TOPLAN and POPLAN, we notice that neither is a clear winner. While TOPLAN has a potentially large branching factor, the search depth is fixed at N, the total number of operators in an optimal plan. POPLAN, on the other hand, has a depth that is P times larger than that of TOPLAN, where P is the maximum number of preconditions for a plan step. In addition, although POPLAN has a bounded branching factor, it is determined not only by the number of existing and new establishers (B_{new} and B_{old}), but also by the number of threats t and the number of ways to resolve a threat r. When the number of preconditions for each plan step and the threats are large, the effective branching factor for partial-order planning could also be extremely large!

Despite the above negative conclusion, the original intuitions of partial-order planning still remain appealing. A partial-order planner works on one subgoal at a time. This is essentially an implementation of the well-known *divide-and-conquer* strategy to problem-solving. One impediment to this implementation is that individual solution plans interact, which in turn increases the number of threats considered (the factor t in time complexity formula).

Another valuable intuition of partial-order planners is that the steps in a plan need not be prematurely sequenced; if most of the steps of a solution plan will eventually be found to be independent of each other, a situation where most plan steps are commutative, then it is indeed better to leave them unordered.

In the rest of the book, we plan to capture the intuitions of *divide-and-conquer* under a different implementation from the one used by basic planning algorithms, an implementation we refer to as *problem decomposition*. In this process, a complex problem is first decomposed into several more-or-less independent parts. Each part is then planned for separately, in accordance with the divide-and-conquer strategy. For each subproblem, a forward-chaining planner or a partial-order planner could be used. If the former, the decomposed domain structure will offer an additional advantage, by reducing the number of applicable operators to each state within each subproblem. This helps reduce the forward-search branching factor. In addition, the solution length for each subproblem is also reduced. This leads to a reduction of search depth in both forward and backward planners. Finally, solutions are combined in such a way that most of the partial-order structures are maintained as much as possible. This policy will most likely lead to increased compactness of solution plans since most of the plan components are commutative. Orders are imposed upon plan steps of solutions to different subproblems only when absolutely necessary. In this manner, *partial-ordering* is a natural by-product of the divide-and-conquer strategy.

In addition to the divide-and-conquer strategy, planning at different *levels of importance* is another often-used approach. Distinguishing between important and unimportant open preconditions and threats entails that at any time a planner need only work on a small set of tasks, and that what has been achieved at higher levels of abstraction could be usefully employed to constrain search at lower levels. For both forward and backward chaining planners, this method of *hierarchical abstraction* could lead to reduced branching factors and depths.

Based on the above motivation, in Parts II and III we will present planning algorithms that are based on divide-and-conquer and hierarchical abstraction and approximation in two separate parts.

3.6 Background

Early work on analyzing the time complexity of forward-chaining, total-order planners was done by Korf [83]. Starting in late 1980s researchers began to compare different versions of backward-chaining planners, in particular partial-order versus total-order planners. Minton, Drummond, and Bresina

[95] presented one such analysis, showing that an instance of a partial-order planner can be more efficient than its backward-chaining, total-order counterpart, although not necessarily always so. Barrett and Weld [17] gave an in-depth comparison of a wide variety of partial-order planners and total-order planners, both being backward-chaining ones, demonstrating that in general the former is more efficient than the latter.

Work has also been done on comparing versions of partial-order planners. When SNLP was first presented at the AAAI91 conference, McAllister claimed that it was more efficient than a previous planner TWEAK, because, due to its use of causal-link protection, it would never generate the same plan twice. This absolute efficiency claim turned out to be flawed, as was shown later analytically and empirically by two independent groups, Knoblock and Yang [81] and Kambhampati [72]. These two works were later combined into a joint paper [74]. The analysis for partial-order planners in this chapter is adapted from a paper co-authored with Craig Knoblock [81], which won the *best paper award* at the 1994 Canadian AI Conference.

The cautious protection of causal-links did have a very important side effect, which turned out to be the cornerstone for subsequent representational extensions of planning domain languages, mainly developed at University of Washington, Seattle. The side effect is that due to the cautious protection of causal links by SNLP, the amount of ambiguity within a single plan is drastically reduced. For example, for every precondition it is clear which plan step is responsible for achieving it. Work in this direction includes [105], [106], [85], and [45].

Bacchus and Kabanza performed an empirical comparison of a forward-chaining, total-order planner and backward-chaining, partial-order planners [10]. They argued that with the former it is much easier to encode domain knowledge for pruning large portions of the search space. This work is predated by work done by Currie and Drummond [39], who designed a method for controlling search with a backward-chaining, partial-order planner. It was reported in [157] that with partial-order planners, Currie and Drummond's method is not always a clear winner. Peot and Smith [107] and Joslin and Pollack [68] presented thorough evaluations of the effects of delaying the removal of threats on planning efficiency. The quest for better heuristics for guiding partial-order planners is ongoing.

4. Useful Supporting Algorithms

4.1 Propositional Unification Algorithm

A key component of a planning algorithm is deciding whether two literals l_1 and l_2 can be made identical. Recall that a literal is either a proposition such as $p(t_1, t_2, \ldots, t_n)$ or its negation. Each parameter t_i, known as a term, of a literal can either be a variable $?x_i$ or a constant C_i. The literals $p(?x,\ C_1)$ and $p(C_2,\ ?y)$ can be made identical by *substituting* the variable $?x$ by C_2 and $?y$ by C_1. However, the literals $p(?x,\ C_1)$ and $q(?u)$ cannot be made identical, nor can $p(?x,\ C_1)$ and $\neg p(?x,\ C_1)$.

Substitution. If two literals can be made identical, they are called *unifiable.* To identify whether two literals are unifiable requires the introduction of *substitution.* A substitution can be defined as a set of pairs

$$\theta = \{\langle ?x_1, t_1\rangle, \langle ?x_2, t_2\rangle, \ldots, \langle ?x_n, t_n\rangle\}$$

where each $?x_i$ is a variable, and each term t_i is either a variable or a constant.

By *applying* a substitution θ to a literal l we mean that every occurrence of $?x_i$ in l is *simultaneously* replaced by t_i, for $i = 1, 2, \ldots n$. The word "simultaneous" refers to the requirement that each replacement is done only once, the replacements for different terms are separately done, and the substitution process is *not* recursive. The result is denoted by $l\theta$. For example, for a literal $l = p(?x, ?y, ?z)$ and substitution $\theta = \{\langle ?x, ?z\rangle, \langle ?y, ?x\rangle, \langle ?z, C\rangle\}$, $l\theta = p(?z, ?x, C)$. $l\theta$ is called an *instance* of l.

Composition. Two substitutions can be composed together.
Let $\theta = \{\langle ?x_i, t_i\rangle, i = 1, \ldots n\}$ and $\sigma = \{\langle ?y_j, u_j\rangle, j = 1, \ldots k\}$ be two substitutions. COMPOSITION(θ, σ) denotes a substitution γ such that for any literal l, $(l\theta)\sigma = l\gamma$. The composition γ is produced by the following steps:

a) produce a substitution $\gamma_1 = \{\langle ?x_i, t_i\sigma\rangle, i = 1, \ldots n\}$, by applying σ_i to every term t_i;
b) remove all identities of the form $\langle ?x_i, ?x_i\rangle$ from γ_1, giving rise to γ_2;
c) for each variable y_j, $j = 1, \ldots k$, if y_j does not appear in $\{x_i, i = 1, 2, \ldots n\}$, add $\langle ?y_j, u_j\rangle$ to γ_2;
d) return the resultant substitution γ.

Unification. For two literals, l_1 and l_2, there may be a substitution θ such that $l_1\theta$ is identical to $l_2\theta$. θ is called a unifier of l_1 and l_2. If no such θ exists, then we say that l_1 and l_2 cannot be unified.

We consider unification in the context of a plan. Let Π be a plan; Π consists of a set of constraints on variable binding. l_1 and l_2 can be unified only when these constraints are not violated.

To find a unifier for the two literals, we could first test to see if they have the same number of parameters and the same propositional symbol (that is, the symbol p in $p(?x)$). If these tests are passed, we could then scan the two literals from left to right, building a unifier θ along the way. Initially, θ is an empty set. At the ith step, the ith parameters of l_1 and l_2 are simultaneously extracted. Let them be u and v respectively. u and v could fall under one of the following cases:

Case 1: u and v are identical. In this case we move on to examine the next pair of parameters.

Case 2: u and v are different constants. In this case the two literals cannot be unified. The unification algorithm returns Fail.

Case 3: At least one of u and v is a variable, and u and v are constrained in the plan Π by a separation constraint: $u \neq v$. In this case l_1 and l_2 cannot be unified either, and the algorithm returns Fail.

Case 4: At least one of u and v is a variable, and u and v are *not* constrained by a separation constraint. Let u be the variable. Form a substitution $\{\langle u, v\rangle\}$. θ is updated by

$$\theta := \textsc{Composition}(\theta, \{\langle u, v\rangle\}$$

The algorithm P-UNIFY is fully described in Table 4.1.

Instantiating Steps, Operators and Plans. Unification is often performed when an open precondition is achieved. Let p be a precondition literal of a plan step s_{need}. Let e be an effect literal of an operator or another plan step $\mathbf{s}$. Suppose that p and e are unifiable, and let θ be P-UNIFY(p, e, Π). An instance of the plan can then be obtained by applying θ to all elements of the plan Π. This means that for all pairs $\langle ?x, t\rangle$ in θ, for all occurrences of $?x$ in Π, including plan steps, causal links, and separation constraints, $?x$ is replaced by t. If we decide to add the new step $\mathbf{s}$ to the plan, we must first apply θ to all elements of $\mathbf{s}$ as well. The resultant plan step is called an *instance* of the originating step or operator. A plan in which all variables are substituted by constants is called a *fully instantiated plan.*

In a plan, all variable binding constraints of the form $?x = ?y$, forcing $?x$ and $?y$ to be bound to the same constant, are implemented as actual substitutions: all occurrences of $?x$ are replaced by $?y$, or vice versa. Thus, the only type of variable-binding constraints actually maintained in a plan is separation constraints.

Table 4.1. The POPLAN algorithm

Algorithm P-UNIFY(l_1, l_2, Π);
Input: Two literals l_1 and l_2, and a plan Π;
Output: A unifier θ if l_1 and l_2 are unifiable. Fail otherwise.

```
1    if one of l1 and l2 is negated and the other is not, or
       l1 and l2 have different propositional symbols, or
       l1 and l2 have different number of parameters
2        then return(Fail);
3    θ := ∅;
4    let n be the number of parameters in l1;
5    for i = 1 to n do
6      let u be the i^th parameter of l1,
7      let v be the i^th parameter of l2;
8      if u ≠ v then
9        if one of u and v is a variable, (say, u), and
           (u ≠ v) ∉ Binding(Π)
         then
10         θ := Composition(θ, {⟨u, v⟩});
11         l1 := l1θ; l2 := l2θ;
12         if l1 = l2
13           then return θ;
14         end if (step 12);
15       else
16         return(Fail);
17       end if (step 9)
18     end if (step 8)
19   end for (step 5)
20   return(θ);
```

4.2 Graph Algorithms

In the process of developing a plan, many queries regarding ordering constraints need be answered. These queries arise in a variety of circumstances; each of them requires a supporting graph-connectivity algorithm:

Cycle Detection: is the plan a true partial-order? That is, is there a cycle in the plan?

Precedence Relation: find the set of all steps which could be ordered before a certain step, without violating the partial-order relation.

Topological Sort: generate a total-order consistent with a partial-order.

In addition, later we describe two algorithms, ALPINE [80] and HIPOINT [13], for generating abstraction hierarchies. These algorithms depend on a graph algorithm for detecting *strongly connected components*. A strongly connected

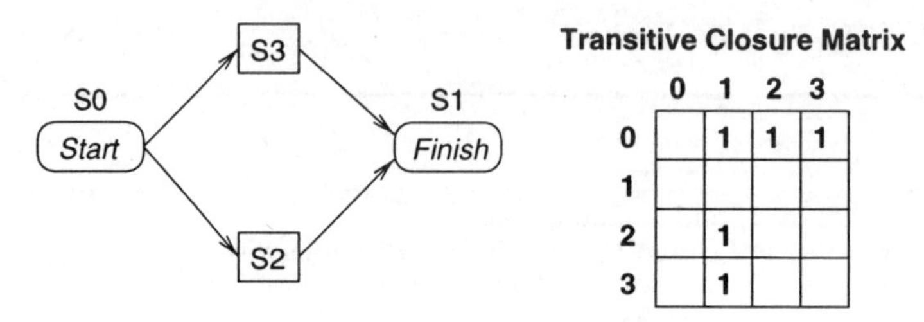

	0	1	2	3
0		1	1	1
1				
2		1		
3		1		

Fig. 4.1. A plan and its associated transitive closure matrix

component in a graph is a set of vertices such that for every pair of vertices u and v, there is a path from u to v and one from v to u.

In this section, we review relevant graph connectivity algorithms to handle these queries. We assume that all steps in a plan Π are uniquely numbered such that the numbering for a step $\mathbf{s}_i$ is i. We also assume that the ordering information is represented using a matrix C^*. For any two steps $\mathbf{s}_i$ and $\mathbf{s}_j$, $C^*(i,j) = 1$ if $\mathbf{s}_i \mapsto \mathbf{s}_j$ in plan Π. Otherwise $C^*(i,j) = 0$. C^* is called a *transitive closure.* An example plan and its corresponding transitive-closure matrix are shown in Figure 4.1.

Suppose that a plan Π has $n-1$ steps, and C is the $(n-1) \times (n-1)$ matrix representing the step-connectivity information. Now suppose that we added a new step $\mathbf{s}_n$. C could be updated by adding a new n^{th} row and a new n^{th} column, such that $C(i,n) = 1$ if $\mathbf{s}_i \mapsto \mathbf{s}_n$ and $C(n,i) = 1$ if $\mathbf{s}_n \mapsto \mathbf{s}_i$. The transitive closure could then be obtained by executing WARSHALL's algorithm (see Table 4.2).

With the transitive closure matrix C^*, various queries could be easily answered:

Cycle Detection: The plan steps form a cycle if for some i, $C^*(i,i) = 1$.
Precedence Relation: A step $\mathbf{s}_i$ is before $\mathbf{s}_j$ in a plan if and only if $C^*(i,j) = 1$. Likewise, an ordering constraint $\mathbf{s}_i \mapsto \mathbf{s}_j$ is consistent with the existing partial order of a plan if $C^*(j,i) = 0$.

Based on the connectivity matrix , we can also implement a depth-first search algorithm, as shown in Table 4.3. This algorithm performs a depth-first search starting with a vertex i. We associate each vertex with a mark, which initially is set to zero. We then systematically scan the vertices beginning with i. To perform depth-first search on a plan, we can place a function call DEPTHFIRST(0) where 0 is the index of the start step $\mathbf{s}_{init}$.

With the depth-first search algorithm, topological search can be easily implemented, whereby a total order of plan steps consistent with the existing partial order can be generated. Right after each step-index i is marked (step 1, Table 4.3), we can output i. The total order is the sequence of output step indexes.

Table 4.2. WARSHALL's algorithm

Algorithm WARSHALL(C)
Input: A connectivity matrix C.
Output: A transitive closure matrix C^*.

```
1.   C* := C;
2.   for k := 1 to n do
3.     for i := 1 to n do
4.       for j := 1 to n do
5.         if C*(i,j) = 0 then
6.           C*(i,j) := C*(i,k) * C*(k,j);
7.         end if
8.       end for
9.     end for
10.  end for
11.  return (C*);
```

Table 4.3. A depth-first search algorithm

Algorithm DEPTHFIRST(i);
Input: A vertex index i, a connectivity matrix C.
Output: A depth-first search of vertices starting with vertex i.

```
1.   Mark(i) := 1; /* i is the vertex being visited */
2.   let Children be the set of indexes j such that C(i,j) = 1 and
     that there is no such k with C(i,k) = 1, C(k,j) = 1;
3.   for each j ∈ Children do
4.     if Mark(j) = 0 then
5.       DEPTHFIRST(j);
6.     end if
7.   end for
```

Similarly, for a directed graph with cycles (not a graph of plan steps), if we have a transitive closure C, we can find out all strongly connected components by looking for indexes i and j such that $C(i,j) = C(j,i) = 1$, and putting i, j in the same component.

Aho et al. [2] present algorithms for depth-first search, topological sort and finding strongly connected components based on a linked-list representation for graph connectivity. Under this representation, the algorithms can be more efficient, although building and updating the list requires additional computational overhead.

4.3 Dynamic Programming

A frequently encountered problem in planning is finding a sequence of decisions subject to certain criteria. Each decision represents a choice made among a set of alternatives. In plan generation, a decision can be the selection of a next action to perform, a choice of a threat resolution method, or a way to establish an open precondition. Because the impact of an individual decision might not be apparent on the overall cost of a sequence of decisions, different sequences need be compared in an efficient manner. Dynamic programming offers a very useful tool for managing this task.

The aim of dynamic programming is to find an optimal sequence of decisions that leads to a goal, where optimality is defined by the user. Consider a set of possible decisions $D = \{d_1, d_2, \ldots, d_n\}$ that an agent might take. Consider also a space consisting of all possible sequences and subsequences of decisions drawn from $\mathcal{P}(D)$, the power set of D. Each sequence in this space leads to a state. From a state S, one can obtain another state T by choosing a next decision d and adding d to S. Also, suppose that attached to each decision $d(S,T)$ that leads from S to T there is a cost value $\mathrm{Cost}(d(S,T))$. The states and their connectivity can be *implicitly* represented as a state-space graph (see Figure 4.2).

The core of dynamic programming method is the *principle of optimality*. Let $S_1, S_2, \ldots, S_n$ be the set of all states directly connected to state T

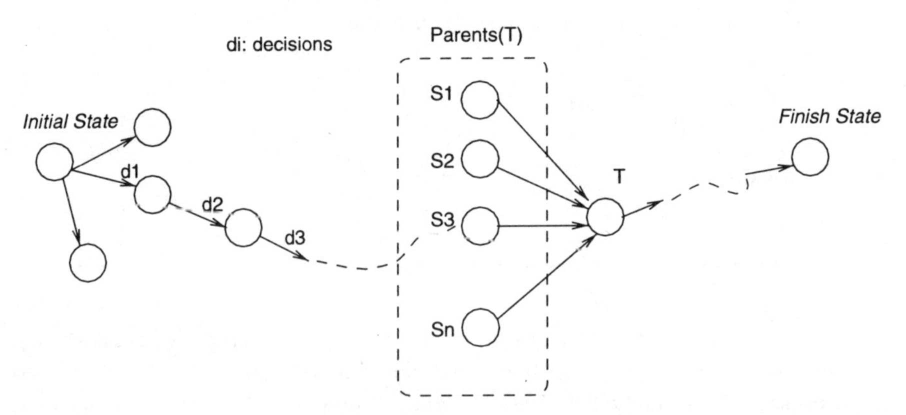

Fig. 4.2. An illustration of dynamic programming algorithms

Table 4.4. The DYNAMIC algorithm

Algorithm DYNAMIC(T)
Input: A state T.
Output: A cost value Cost and a set of optimal decisions Opt-Decisions(T), leading to state T.

1. **if** T is a start state **then**
2. **return**(0);
3. **else**
4. Parent-List := Parents(T);
5. let S be a parent of T in Parent-List such that Cost := (Cost($d(S, T)$) + DYNAMIC(S)) is minimal;
6. Opt-Decisions(T) := Opt-Decisions(S)$\cup\{d(S, T)\}$;
7. **end if**;
8. **return**(Cost);

---

via a single decision $d(S_i, T), i = 1, 2, \ldots n$. The principle of optimality says that the optimal sequence of decisions passing through T must be obtained from the *optimal subsequence* of decisions from the initial state S_{init} (S_{init} is an empty set of decisions) to T, and that from T to the final state. Let MinCost(T) represent the cost of a minimal cost path reaching T from S_{init}. In terms of costs, the principle of optimality translates to the following formula (see Figure 4.2):

$$\text{MinCost}(T) = \min\{\ [\text{MinCost}(S_i) + \text{Cost}(d(S_i, T))]\ \mid i = 1, 2, \ldots n\}$$

The value of this formula is that it is inductive. One can start from an empty set of decisions and recursively compute a minimal-cost decision sequence *incrementally.* In each step of computation, the decision $d(S_i, T)$ which gives rise to a minimal cost value of state T is remembered. When a final state is reached, the optimal sequence of decisions can then be traced backwards from the goal.

In Table 4.4 we list an implementation of a dynamic programming algorithm using recursion. It is assumed that for the problem at hand, there is a start state and a final state. The aim is to find an optimal path from the start state to the final state via a sequence of optimal decisions. To apply this algorithm, one needs a supporting subroutine Parents(S) (see Figure 4.2) for generating a set of parent states from any given state. The algorithm can be activated by a program call with a final state as an input parameter.

Dynamic programming is often applied to *explicit graphs.* A state can be formulated by either a single node in a graph or a set of nodes. An advantage of *dynamic programming* is that with the explicit graph as input, the states on which computation is made need only be *implicitly* represented. After each step of computation only the optimal decision is remembered (step 6 in Table 4.4), and the rest of the state connections forgotten. Because of this feature, dynamic programming can function rather effectively even when the set of states S_i leading to any particular state T is large.

In Chapter 8 we consider how to *merge* plans using an implementation of a dynamic programming algorithm.

4.4 Branch and Bound

Dynamic programming is a computational method for choosing among sequences of alternative decisions. Another useful alternative-selection method is branch-and-bound. Like dynamic programming, this method depends on modeling a problem as a state-space search problem. A state can be a configuration of the world much in the same way as the states in the planning algorithm ToPlan, or it can be a set of chosen decisions as in the dynamic programming model. It is assumed that given any state S, a function Successors(S) generates the set of all children states from S. These children states are then entered in some order into a list of nodes known as an *active list*. In each iteration of a branch and bound algorithm, a state is chosen to be expanded, producing its successor states. This state is called an *entering state*.

The aim of branch and bound is to find a minimum cost goal state. What makes branching and bound different from the other algorithms is its ability to trim the active list by using two estimated cost values, a *lower bound* and an *upper bound* cost value for each state. The upper bound $U(S)$ for a state S is a current estimate on how high the cost values of S as well as all possible descendents of S can reach. If T is a descendent of S, then $\text{Cost}(T) \leq U(S)$. Similarly, a lower bound value $L(S)$ estimates how low the cost values of S and the descendents of S can be. For every state S, $L(S) \leq U(S)$. Furthermore, it is assumed that for a goal state, we know its cost value exactly. If G is a goal, then $L(G) = U(G) = \text{Cost}(G)$. It is also assumed that for every state, these two estimates are provided by the user.

Suppose that the active list contains two states S_1 and S_2. If $L(S_1) > U(S_2)$ then all descendents of S_1 must have a higher cost than any descendent of S_2. This indicates that S_1 can be removed from the active list without losing a least-cost solution. This process is shown in Figure 4.3.

An algorithm for branch and bound is shown in Table 4.5.

Given the similar goals for both dynamic programming and branch-and-bound, how to we choose between these two methods for a given application? The "branching" part of a branch and bound algorithm corresponds to the generation of children states from a parent state. To make this process feasible, branch and bound is usually applied when the number of children states is small. In addition, we must have accurate estimates for the upper and lower bounds; the tighter the two bounds, the more the number of states pruned.

Table 4.5. The BRANCH-AND-BOUND algorithm

Algorithm BRANCH-AND-BOUND(S)
Input: An initial state S
Output: An optimal goal state G.

```
1.    A := {S};
      /* (A is the branch-and-bound active set)*/
2.    repeat
3.      remove from A a state S for which L(S) is smallest;
4.      if S is a goal state then return(S)
5.      else
6.        C := Successors(S);
7.        for each node S1 in A do
8.          for each node S2 in C do
9.          if L(S1) > U(S2) then
10.           remove S1 from A;
11.         else if L(S2) > U(S1) then
12.           remove S2 from C;
13.         end if
14.       A := A ∪ C;
15.     end if
16.   until A = ∅;
17.   return(Fail);
```

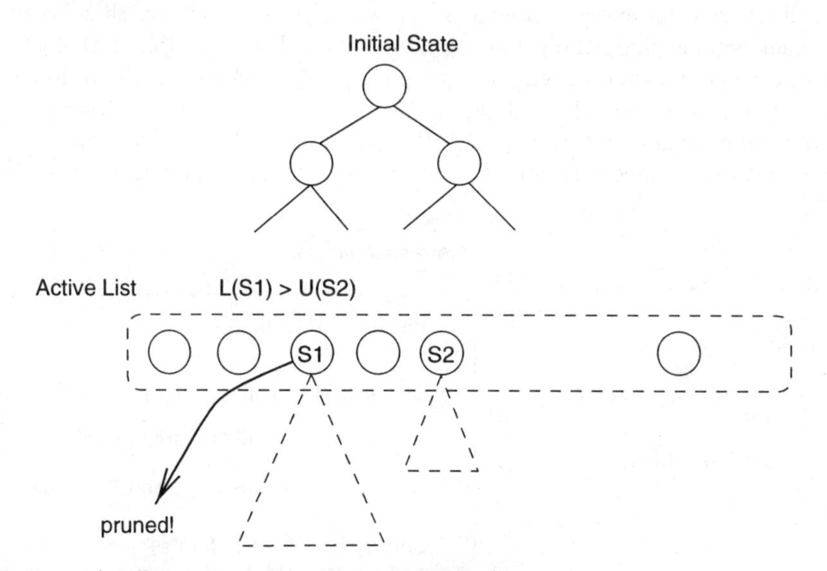

Fig. 4.3. An illustration of branch and bound algorithms

When the number of successors of any state is large, dynamic programming is recommended because of its ability to compress all successors generated from a given state by an optimal decision state.

4.5 Constraint Satisfaction

Constraint satisfaction problems (CSPs) provide a simple and yet powerful framework for solving a large variety of AI problems. The technique has been successfully applied to machine vision, belief maintenance, scheduling, and many design tasks. In Chapters 5 and 7 we will describe how it is used for modeling and solving planning problems.

A constraint satisfaction problem (CSP) can be formulated abstractly as three components:

1. a set of *variables*, $X_i, i = 1, 2 \ldots n$,
2. for each variable X_i a set of values $\{v_{i1}, v_{i2}, \ldots v_{ik}\}$. Each set is called a *domain* for the corresponding variable, denoted as Domain(X_i),
3. a collection of *constraints* that defines the permissible subsets of values to variables.

The goal of a CSP is to find one (or all) assignment of values to the variables such that no constraints are violated. Each assignment, $\{x_i = v_{ij_i}, \ i = 1, 2, \ldots, n\}$, is called a *solution* to the CSP.

As an example of a CSP, consider a map-coloring problem, where the variables are regions $R_i, i = 1, 2, \ldots, n$ that are to be colored (see Figure 4.4). In a final solution every region must be assigned a color such that no two adjacent regions share the same color. A domain for a variable is the set of alternative colors that a region can be painted with. For example, a domain for R_3 might be {Green, Red, Blue}. Between every pair of adjacent variables, a constraint exists that states that the pair cannot be assigned the same color. Between adjacent regions R_1 and R_2, for example, there is a constraint

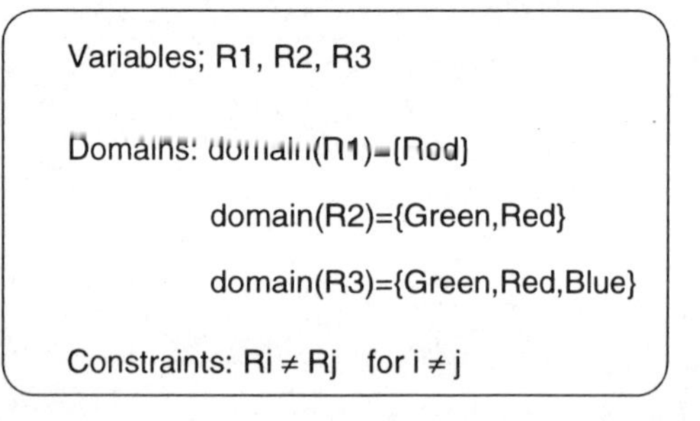

Fig. 4.4. A CSP representation of the graph coloring problem

$R_1 \neq R_2$, stating that a value assigned to R_1 must be different from one assigned to R_2. A solution to the problem is a set of colors, one for each region, that satisfies the constraints.

Let VARS = $\{X, Y, \ldots Z\}$ be a set of variables. A constraint on VARS is essentially a *relation* on the domains of the variables in VARS. If a constraint relates only two variables then it is called a *binary constraint.* A CSP is binary if all constraints are binary. For any two variables X and Y, we say $X = u$ and $Y = v$ are *consistent* if all constraints between X and Y are satisfied by this assignment.

A variety of techniques have been developed for solving CSPs. They can be classified as local *consistency-based methods*, global *backtrack-based methods*, or *local-search methods.* Local-search methods is a kind of greedy algorithm which is gaining popularity. In the last section we review a special case of local search algorithms for solving large-scale satisfiability problems. An application of local search to scheduling can be found in a paper by Minton et al. [96].

4.5.1 Local-Consistency Based Methods

Local consistency methods follow the theme of *preprocessing.* That is, before a more costly method is used, a consistency-based method can be applied to simplify a CSP and to remove any obviously incompatible values. Often these methods yield tremendous headway toward eventually solving the problem.

Let X and Y be two variables. If a domain value A of X is *inconsistent* with all values of Y, then A cannot be part of a final solution to the CSP. This is because in any final solution S, any assignment to X must satisfy all constraints in the CSP. Since $X = A$ violates at least one constraint in all possible solutions, A can be removed from the domain of X without affecting any solution. The procedure is implemented in a subroutine REVISE (see Table 4.6).

Figure 4.5 illustrates the operation of REVISE. In this example, the value `A` of a variable X is inconsistent with every value of variable Y; `A` is therefore removed from the domain of X without losing any potential solutions. In the graph-coloring example of Figure 4.4, this procedure removes the value `Red` from the domain of region `R3`, making the problem simpler to solve.

If for a pair of variables (X, Y), for every value of X there is a corresponding *consistent* value of Y, then we say (X, Y) is arc-consistent. By the above argument, enforcing arc-consistency by removing values from variable domains does not affect the final solution. The condition that every pair of variables is arc-consistent is called *arc-consistency.*

The algorithm AC, described in Table 4.6, enforces arc-consistency for all pairs of variables (X, Y). It is assumed that for any CSP, Arcs(CSP) gives the set of all pairs of variables which are constrained under a binary constraint. The algorithm applies a subroutine REVISE to every constrained pair of variables (X, Y), to ensure the the pair is arc-consistent. If not, REVISE will

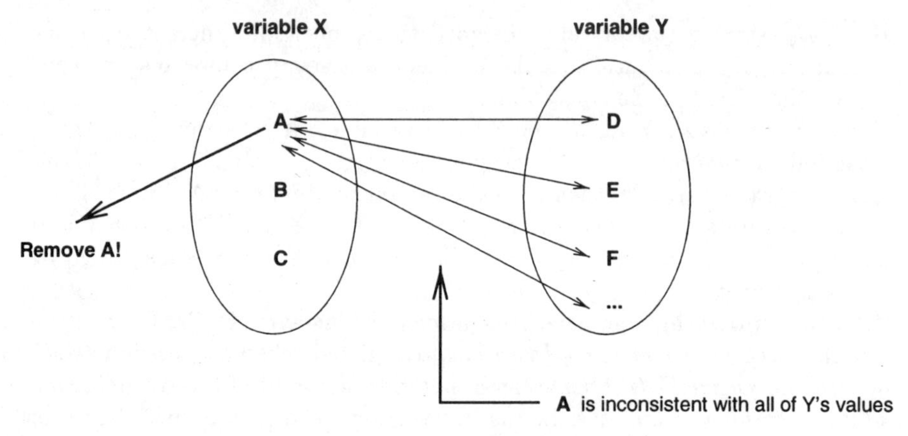

Fig. 4.5. Revising the domain of a variable in a CSP

remove inconsistent values from the domain of X. Since the removal of certain values from Domain(X) might prompt new violations of arc-consistency for all variable pairs $(Z, X) \in$ Arcs(CSP), these pairs are added back to the queue (Step 6) by AC for re-examination.

4.5.2 Backtrack-Free Algorithm

In some special cases, it is even possible to solve a CSP entirely by studying its graph topology. In this graph every variable is a node, and two nodes are connected by an arc if there is a constraint between them. Freuder [53] gave the following theorem:

Theorem 4.5.1 (Freuder 82)
If (1) the topology of the CSP is a tree, (2) the consistency relations between variables are binary, and (3) arc-consistency is satisfied by the graph, then the CSP can be solved without backtracking.

To see this, suppose that a CSP satisfies the three conditions in the above theorem. It can be solved by selecting a value from the root node and repeatedly finding a consistent value in a next adjacent node. The procedure repeats until a value is selected for every node. Because all relations are binary and the graph is a tree, the values in the next adjacent node only interact with the values in the immediately previous node. Because arc-consistency is satisfied, any partial solution selected is guaranteed to be consistent with the next node. Thus, it is possible to extend the current partial solution by including one more variable-value pair. By induction, the entire CSP can be solved without backtracking.

Table 4.6. The AC algorithm

Algorithm AC(CSP)
Input: A CSP
Output: An arc-consistent CSP;

```
1.    Queue := {(X, Y) | (X, Y) ∈ Arcs(CSP)};
2.    loop
3.      select and remove (X, Y) from Queue;
4.      if REVISE(X, Y) = TRUE then
5.        affected-pairs := {(Z, X) | (Z, X) ∈ Arcs(CSP)};
6.        Queue := Queue∪affected-pairs;
7.      end if
8.    until Queue = ∅
9.    return(CSP);
```

Subroutine REVISE(X,Y)
Input: A pair of variables X and Y
Output: TRUE if the Domain(X) is revised; FALSE otherwise.

```
1.    Delete := FALSE;
2.    for each u in Domain(X) do
3.      REMOVEFLAG:= TRUE; /* if false we know some value is removed */
4.      for each v in Domain(Y) do
5.        if X = u, Y = v are consistent, then
6.          REMOVEFLAG := FALSE;
7.        end if
8     end for;
9.      if REMOVEFLAG = TRUE then
10.       Domain(X) := Domain(X) − {u};
11.       Delete := TRUE;
12.     end if
13.   end for;
14.   return(Delete);
```

4.5.3 Backtrack-Based Algorithms

Arc-consistency algorithms only work on pairs of variables, and as such can only handle binary constraints and cannot always guarantee a final solution to a CSP. A more thorough method for solving a CSP is backtracking, where a depth-first search is performed on a search tree formed by the variables in the CSP.

A backtracking algorithm instantiates the variables one at a time in a depth-first manner. It backtracks when the constraints accumulated so far

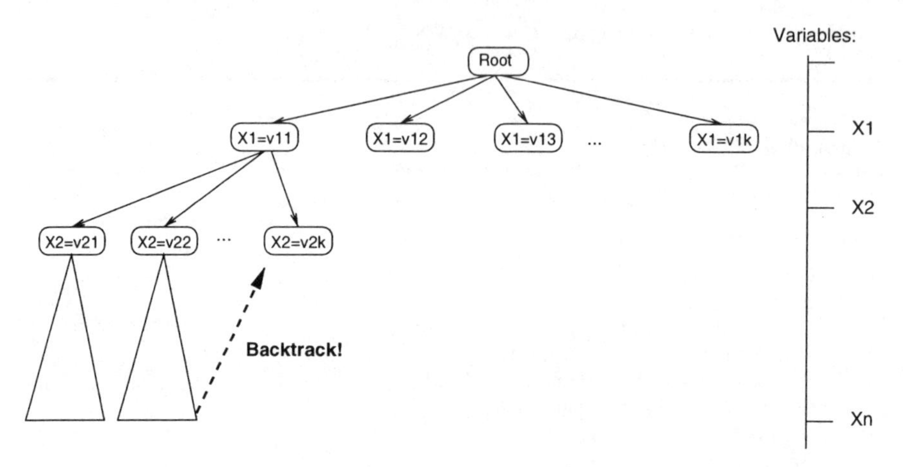

Fig. 4.6. A search tree for a backtrack-based algorithm

signal inconsistency. In Figure 4.6 we show this process. First, variables are ordered in a certain sequence. Different orders of variables might entail different search efficiency, and heuristics for good ordering of variables are called *variable-ordering heuristics*. Similarly, for each variable, the values are tried out one at a time, and the heuristics for a good ordering of values are called *value-ordering heuristics*.

We focus on how to combine backtracking and arc-consistency to yield a more intelligent backtrack-based algorithm. As shown in the figure, the backtracking algorithm searches in a space of nodes. A node consists of a list of assignments already made to variables and a list of remaining variables yet to be assigned values:

$$N = \langle \texttt{Assigned}, \texttt{Rem} \rangle$$

where $\texttt{Assigned} = \{X1 = v_1, X2 = v_2, \ldots, X_m = v_m\}$ is the current assignment, and Rem is a set of variable/domain pairs for all remaining variables:

$$\begin{aligned} \texttt{Rem} \quad = \{ \quad & (X_{m+1}, \{v_{m+1,1}, v_{m+1,2}, \ldots\}) \\ & (X_{m+2}, \{v_{m+2,1}, v_{m+2,2}, \ldots\}) \\ & \ldots \\ & (X_n, \{v_{n,1}, v_{n,2}, \ldots\}) \end{aligned}$$

During the backtracking search, the children nodes of N can be formed by assigning one of the values $v_{m+1,j}$ to the first remaining variable X_{m+1}, resulting in a new node:

$$N_j = \langle \texttt{Assigned} \cup \{(X_{m+1} = v_j)\},\ \mathrm{Rest}(\texttt{Rem}) \rangle$$

where the function Rest returns a list with the first element removed.

For each child N_j, partial arc-consistency can be enforced on the new set of remaining variables: for every remaining variable and its associated

Table 4.7. The FwdChecking algorithm

Algorithm FwdChecking(CSP)
Input: A CSP: CSP $= \{(X_i, \text{Domain}(X_i)) \mid i = 1, 2 \dots n\}$
External Functions:
FwdRevise((X, v), Rem) updates the variable domains by a new assignment $X = v$;
VarOrdering(CSP) implements a variable-ordering heuristic;
ValOrdering(Valset) implements a value-ordering heuristic;
Output: A solution to the CSP if a solution exists; Fail otherwise.

```
1.   if CSP = ∅ then
       return(∅);
2.   else
3.     CSP := VarOrdering(CSP);
4.     (X, Valset) := First(CSP);
5.     Valset := ValOrdering(Valset);
6.     for each value v in Valset do
7.       NewA := {(X_{m+1}, v)};
8.       Rem := FwdRevise(NewA, Rest(CSP));
9.       if (SubSol = FwdChecking(Rem)) ≠ Fail then
10.        return(NewA∪SubSol);
11.      end if
12.    end for
13.  end if
14.  return(Fail);
```

domain values $(X_i, \{v_{i,1}, v_{i,2}, \dots\})$, remove all values $v_{i,l}$ for which $(X_{m+1} = v_j, X_i = v_{i,l})$ is inconsistent. By the same reason as with arc-consistency, this value-removal operation does not delete any potential solutions to the CSP. If as a result of this processing some value set becomes empty, then the entire subtree associated with that child should correspond to a dead end. Search need not go further along that direction. This procedure is essentially a version of the subroutine Revise. We will call it FwdRevise.

This simple amendment to backtracking literally checks forward after every node expansion. For this reason it is called **forward-checking**. Forward-checking has been shown to be one of the most efficient methods for solving a CSP. Table 4.7 shows the algorithm.

In the FwdChecking algorithm two heuristic functions are provided by the user. The function VarOrdering(CSP) implements a variable-ordering heuristic for a CSP, and the function ValOrdering(Valset) implements a value-ordering heuristic on a set of values. These heuristics have been tested under several schemes, classified under either *fail-first* or *fail-last*. The former prefers a search-tree branch where it is most likely to terminate without finding a solution, the latter a path that is most likely to succeed. Each of these heuristics has been shown to be effective in specific domains.

In addition to forward checking, there are also other intelligent backtrack-based algorithms. Some examples are *backjumping* and *backmarking*, both used for implementing ways to record reasons for either failures or success during a backtracking search.

4.6 The GSAT Algorithm

In Chapter 2 we discussed the SATPLAN algorithm for planning. This plan-generation algorithm depends on the GSAT algorithm, which we discuss in this section.

The GSAT algorithm is a special type of local-search algorithm, designed to solve large-scale propositional satisfiability problems. Its input is a set of clauses CSet $= \{C_i, i = 0, 1, 2, \ldots, n\}$, representing the conjunction of the clauses. Each clause C_i is a set of literals $\{p_1, p_2, \ldots, p_k\}$, representing, in this case, their disjunction. The set of clauses represents a *conjunctive normal form* (CNF).

In a CNF, each proposition p_i can be either TRUE or FALSE. If any proposition in a clause is set to TRUE, the clause is said to be satisfied. The goal of the propositional satisfiability problem is to set at least one proposition p_i in every clause C_i to TRUE. The resultant solution is known as a *satisfying assignment.*

Local search can be very effective for several constraint-satisfaction problems. Examples of these problems are the N-queens puzzles and the scheduling problems [126, 96]. In the case of propositional satisfiability problems, it is particularly useful. A propositional satisfiability problem is a special kind of constraint satisfaction problem, in which the propositions are the variables, the TRUE/FALSE values are domain values for the variables, and the requirement that all clauses be satisfied corresponds to a global constraint on the variables.

Given a set of clauses CSet, GSAT starts by generating an initial assignment T. This assignment may fail to set all clauses true. In that case, GSAT finds a proposition such that reversing its truth value (from TRUE to FALSE or vice versa) results in the largest increase in the number of satisfied clauses[1]. This proposition is called a *max-satisfying* proposition, and the operation for changing the truth value of p is known as a *flip*. After each flipping operation, a next assignment is generated by GSAT by reversing the truth value of this proposition. GSAT repeats this process until either a satisfying assignment is found, or a pre-assigned upper bound on the number of flips is exceeded.

[1] These are occasional exceptions. For example, it might only require the smallest decrease in the number of failed clauses.

Table 4.8. The GSAT algorithm

Algorithm GSAT(CSet);
Input: A set clauses CSet, two constants MaxTries and MaxFlips.
Output: a satisfying truth assignment of CSet if found.

```
1   for i := 1 to MaxTries do
2     T := a randomly generated truth assignment;
3     for j := 1 to MaxFlips do
4       if T is a satisfying assignment then
5         return T;
6       else
7         let p be a max-satisfying proposition
          (break ties randomly);
          Comment: now flip p...
8         if p := TRUE in T then
9           set p := FALSE in T;
10        else
11          set p := TRUE in T;
12        end if
13      end if (step 4)
14    end for (step 3)
15  end for (step 1)
16  return(Fail);
```

For example, suppose that a CNF CSet consists of two clauses: $T = \{C_1, C_2\}$, where the clauses are

$$C_1 = \{x, y, \neg z\}, \tag{4.1}$$
$$C_2 = \{\neg x, \neg y, \neg z\} \tag{4.2}$$

Assume that an initial assignment T is defined as $x = \text{TRUE}, y = \text{TRUE}, z = \text{TRUE}$. Under this assignment, C_2 is false. In this case, z is a max-satisfying proposition since setting it to FALSE makes both clauses true. In fact, setting z to FALSE gives rise to a satisfying assignment.

A high-level version of the GSAT algorithm is shown in Table 4.8. Two parameters must be determined before the application of the algorithm. One is MaxTries, designating the maximum number of local-search sessions to try before giving up. The second is MaxFlips which gives an upper bound on the number of flips for satisfying propositions during each local-search session. By default, MaxTries and MaxFlips can be set at N and $5N$, respectively, for a problem with N propositions.

A feature of GSAT is that when choosing which proposition to flip, it is sometimes beneficial to allow a flip to occur even when it does not strictly increase the number of satisfied clauses. Also, a certain amount of non-determinism is found to be useful. For example, if there is more than one proposition whose flipping would result in a maximum increase in satisfied clauses, one can select among them at random. In addition, a few variations of the basic GSAT algorithm are found to be very effective. The most important variation on GSAT is known as the "walk" option, where there is a new probability parameter P whose value lies between 0 and 1. On each flip, with probability P, *any* variable that appears in a unsatisfied clause can be flipped. Likewise, with probability $(1-P)$, a max-satisfying variable is flipped. This option is necessary for solving the planning problems discussed in Chapter 2. As an example of the variations of GSAT, see [120].

For several test problems, GSAT dramatically outperforms backtracking-based search algorithms, often resulting in orders of magnitude reduction in search time. There has been much study on what constitutes the hard problems for GSAT. It was discovered that some problems with a certain proposition/clause ratio are the hardest to solve using GSAT as well as other CSP algorithms. See [34] for a discussion.

4.7 Background

Unification algorithms can be found in several AI textbooks, including those by Nilsson [102], Russell and Norvig [112], and Rich and Knight [110]. Graph-traversal algorithms are thoroughly discussed in a book by Aho, Hopcroft, and Ullman [2]. Dynamic programming and branch-and-bound methods can be found in [66].

Since the early 1980s, Constraint Satisfaction has undergone a tremendous growth both in interest and in application. Early foundational work can be found in [92], [91], [53], [37], [36]. Algorithm AC is based on [92]. Forward-checking is described in [62]. Good overviews on this topic can be found in an article by Kumar [84] and a book by Tsang [162]; a collection of papers can be found in a special issue of *Artificial Intelligence Journal* on constraint-directed reasoning [54].

The GSAT algorithm is based on an IJCAI-93 article by Selman and Kautz [118]. There is a large body of work related to local search techniques and propositional satisfaction problems. See [30], [120, 34], [118], [119] for an overview. See [77] for a discussion on how GSAT is applied to planning.

5. Case Study: Collective Resource Reasoning

As a case study of the application of constraint-satisfaction representation, in this chapter we consider how to extend a traditional backward-chaining, partial-order planner to one that could reason about resources. We present an extended-plan language in which each variable is associated with a finite set of domain values. We discuss the associated changes this brings to planning algorithms. In addition, we present empirical results to demonstrate the efficiency gain of the extended planner, and to answer some of the utility questions regarding planning with constraint satisfaction.

5.1 Eager Variable-Binding Commitments

Over the years, researchers have extensively studied different methods of imposing ordering constraints on the steps of a plan. It has been shown that, when appropriately used, a *delayed-commitment* or *least-commitment* strategy in step-ordering could effectively reduce a planner's search space.

In this chapter, we extend the delayed-commitment strategy in step-ordering to *variable binding*, using techniques in solving constraint-satisfaction problems (CSPs). The idea is to represent possible bindings of a variable in a plan *collectively*, and to instantiate a variable to a constant value only when necessary. This approach to reasoning about resources has been adopted by a number of practical planning systems, most notably MOLGEN [128], SIPE [147], GEMPLAN [88], and scheduling systems such as ISIS [50]. What has not been clear from these previous systems is how to represent the resources in a concise mathematical model of CSP, and how the associated constraint reasoning systems can cope with incremental changes in a plan, which are brought up by the insertion of new steps. In addition, there is an open *utility* question on how much constraint processing should be performed *during planning* to best improve the planning efficiency.

To illustrate the variable-binding problem, consider the household domain. Suppose that our goal is to paint a ceiling (see Figure 5.1). One of the last steps in a solution plan is `apply-paint`, shown at the top of the figure, which has an argument `?brush`. When this step is inserted in the plan, `?brush` is yet uncommitted to any particular constant. At this point, we have two choices: either we arbitrarily choose an available constant for

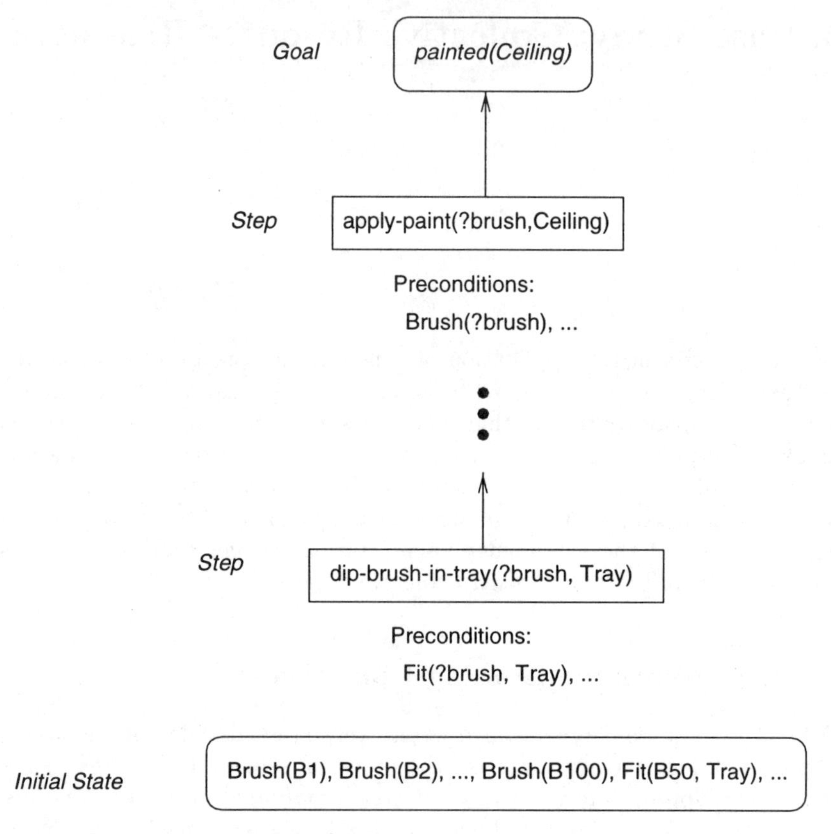

Fig. 5.1. Committing to a paint brush

a brush, and instantiate `?brush` to that constant, or we wait until we have more information before committing to a specific brush.

The first choice could lead to serious computational problems. As shown in Figure 5.2, if there were 100 different brushes, the search tree would split into 100 branches. If, however, another step `dip-brush-in-tray`, for dipping the brush in a paint tray, is inserted into the plan at a later stage, it may be discovered that most of the 100 brushes would not fit into the tray! At this point, many of the branches will correspond to dead ends in the search tree, and at these branches extensive backtracking would occur.

This example exposes a problem with the ad-hoc method of *eager commitment* in variable binding. As shown above, the consequence of eager commitment could make the search tree very bushy, and could lead to a backtracking phenomenon known as *thrashing*. Thrashing occurs in a search process when the same part of the search tree fails repeatedly for the same reason, in this example the reason being that the brushes would not fit into the paint tray.

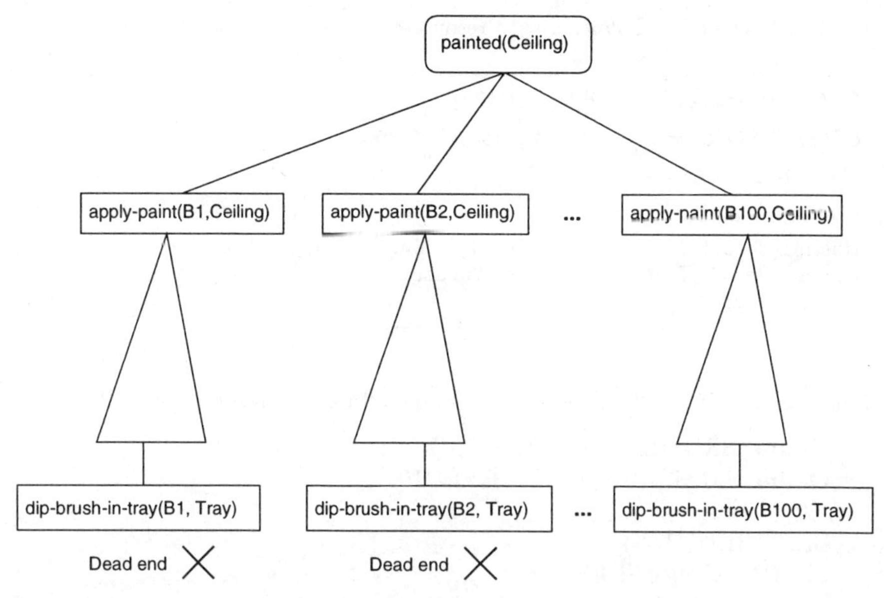

Fig. 5.2. The thrashing problem in search

If this is discovered late in the planning process, a significant amount of computation would be wasted.

A solution to the above problem would be to represent *collectively* the objects to which each variable can be bound. In the next section, we describe a representational method for dealing with this problem.

5.2 Extending the Representation

In our extension of planning domain representation, every variable has a *domain of values*, representing the entire range of values that could be assigned to the variable. For example, in the household domain, a painting operator `apply-paint(?brush`, Ceiling) would be associated with a domain for `?brush`: $\{B1, B2, \ldots, B100\}$. For convenience, we could declare a symbol BrushDom to be this set, and use the representation in Table 5.1 to represent the operator. The denotation of a variable in this extended operator representation is *existential*; for each variable $?x$ in a literal $P(?x)$, the associated variable declaration

$$\textbf{Var: } ?x : \text{VarDomain}$$

stands for

$$\exists\ ?x \in \text{VarDomain}.\ P(?x)$$

In addition, in the initial state representation, all literals sharing the same property are represented succinctly using a set notation. For example, in the

Table 5.1. Operator definition with resources

Define BrushDom: $\{B1, B2, \ldots, B100\}$

OPERATOR: `apply-paint(?brush`, Ceiling)

Var: `?brush`: BrushDom;

Precondition	Effect	Cost
Holding(`?brush`), Dipped(`?brush`, Tray), Can-Reach(Ceiling)	Painted(Ceiling), ¬Dry(`?brush`)	5.0

painting domain, part of the initial state could be represented as follows:

Define BRUSHES: $\{B1, B2, \ldots, B100\}$;
Define SMALLBRUSHES: $\{B50, B100\}$;

Brush(BRUSHES);
Fit(SMALLBRUSHES, Tray)
etc.

In the representation of the initial state, the **Define** part declares a new symbol, which represents a set of objects. The meaning of each literal is *universal*; the literal holds for each of the constants in the set.

A plan Π is represented just as before, except that now each step is associated with a set of variable-domain declarations, just as in an operator definition. The variables in a plan form a constraint satisfaction problem (CSP). In this problem, the set of variables in the steps of a plan are the variables of the CSP, the constraints among variables are either equality or inequality constraints. A solution to the problem is an assignment of values taken from the domains of the variables to the corresponding variables, such that all variable-binding constraints (that is, $?x_1 = ?x_2$, and $?y_1 \neq ?y_2$) are satisfied.

In general, solving this CSP may not be computationally easy. As an example, Figure 5.3 shows a CSP associated with a painting plan. The variables are brushes to be used to paint the objects. Clearly there is no solution to this CSP, but the computation is nontrivial; in general, constraint-satisfaction problems are NP-hard.

Despite the worst-case computational difficulty in solving a CSP, efficient algorithms exist. As mentioned in Chapter 4, the local consistency algorithms and global backtracking-based heuristic algorithms could help detect dead ends early and solve a CSP efficiently. In this chapter, we refer to these CSP algorithms collectively as CSPSOLVE. Given a constraint satisfaction problem CSP, CSPSOLVE(CSP) returns a solution to the problem, where a solution is a consistent assignment to all variables. If the CSP has no solution, CSPSOLVE returns FALSE.

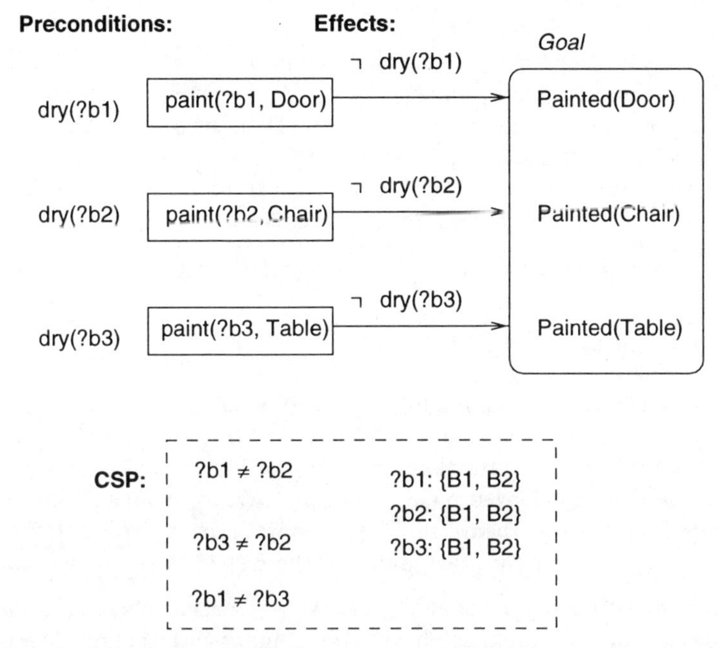

Fig. 5.3. A constraint satisfaction problem associated with a plan

5.3 Plan Generation

Consider a backward-chaining, partial-order planning algorithm. To cope with the new representation, we must extend several key steps to account for constraint reasoning. Readers should consult the backward-chaining, partial-order planning algorithm in Table 2.2. The resultant extended algorithm will be referred to as FDPOP algorithm (Finite-Domain Partial-Order Planner).

5.3.1 Correctness Check

The correctness-checking step (Step 5) of a partial-order planning algorithm verifies the plan to make sure that every precondition of every step has a causal link, and that the causal link has no negative threats. To extend to the CSP representation, all we need add is that in addition to the above check, Step 5 will also verify if a consistent solution to the associated CSP exists. This can always be done using a backtracking algorithm. If so then a correct and consistent instance of the plan Instance(Π) will be returned.

5.3.2 Threat Detection

Threat detection is needed in POPLAN at step 7. The modification for constraint-based threat-detection is simple: for each causal link $\langle \mathbf{s}_i \stackrel{p(x)}{\rightarrow} \mathbf{s}_j \rangle$, a check is made to see if a step $\mathbf{s}_k$ exists such that

a. $\mathbf{s}_k$ could be ordered between $\mathbf{s}_i$ and $\mathbf{s}_j$,
b. $\mathbf{s}_k$ has an effect $\neg p(y)$, such that x and y are not constrainted by a separation constraint, and
c. if x and y are variables, Domain$(x) \cap$ Domain$(y) \neq \emptyset$.
d. If both x and y are constants, $x = y$.
e. Otherwise, if one of x and y is a constant C and the other a variable, C is in the domain of the variable.

Item **c** above indicates that $\mathbf{s}_k$ is a possible threatening step.

5.3.3 Precondition Establishment

The next extension to the planning algorithm is in the `Establish-Precondition` step (Step 10). Let $\mathbf{s}_j$ be a plan step with an open precondition $p(?x)$, where the domain of the variable $?x$ is Domain$(?x)$. This precondition can be achieved by an operator in $\mathcal{O}$, or by an existing step. Of all existing steps, one of them could be the start step, representing the initial state. Here we consider the start step and the rest of the steps separately.

Case 1: Start Step. Suppose that the start step asserts an effect literal $p(\texttt{Set})$, and that `Set` and Domain$(?x)$ have a non-empty intersection. We can let Domain$(?x)$ be this intersection, and we can produce a new causal link $\langle \mathbf{s}_{init} \overset{p(?x)}{\rightarrow} \mathbf{s}_j \rangle$ in the plan.

Case 2: Other Steps and New Operators. Let $\mathbf{s}_i$ be an existing step with an effect literal $p(?y)$. If $\mathbf{s}_i$ could be ordered before $\mathbf{s}_j$ without violating ordering constraints, we would like to make $\langle \mathbf{s}_i \overset{p(?x)}{\rightarrow} \mathbf{s}_j \rangle$ a new causal link in the plan. In doing so, we would like to make sure that $?x$ and $?y$ have a non-empty domain intersection. If this is true, we let the domains of both $?x$ and $?y$ be the same as their intersection. The procedure for inserting a new operator to achieve a precondition requires a similar modification.

The FDPOP algorithm is shown in Table 5.2.

5.4 A House-Painting Example

We now illustrate the FDPOP algorithm with the painting example introduced in Chapter 2 (Section 2.3.1), where our goals are to paint a ceiling, a door and a table. To achieve these goals using the new operator definitions, suppose also that we re-define the operators in Table 2.3, resulting in Table 5.3. Each painting operator has a list of possible brushes that could be used for painting an object. The brushes are collectively represented as variables.

The first several planning operations are basically identical to those described in Section 2.3.4. We focus on the last step, shown in Figure 5.4. In

Table 5.2. The FdPop algorithm

Algorithm FdPop($\mathcal{O}$, Π_{init});
Input: A set of planning operators $\mathcal{O}$, and an initial plan Π_{init} consisting of a start step and a finish step;
Output: A correct plan if a solution can be found.

```
1   OpenList := { Πinit}.
2   repeat
3     Π:= lowest cost plan in OpenList;
4     remove Π from OpenList;
5     if Correct(Π) then
6       if CspSolve(Π) then
7         return(Instance(Π));
8       end if
9     else
10      if Threats-Exist(Π) then
11        Let t be a threat; Succ := Resolve-Threat(t, Π);
12      else
13        Succ := Establish-Precondition(Π);
14      end if
15      Decide whether to apply CspSolve to all successors in Succ;
        If yes, then apply;
16      Add all remaining successors in Succ to OpenList;
17    end if
18  until OpenList is empty;
19  return(Fail);
```

Table 5.3. Operator definition with resources

`paint-ceiling`(?cbr, Ceiling) **Var:** ?cbr: {B1, B2, B3}		
Preconditions	Effects	Cost
Dry(?cbr)	Painted(Ceiling), ¬Dry(?cbr)	5.0
`paint-door`(?dbr, Door) **Var:** ?dbr: {B1, B2}		
Preconditions	Effects	Cost
Dry(?dbr)	Painted(Door), ¬Dry(?dbr)	5.0
`paint-table`(?tbr, Table) **Var:** ?tbr: {B1}		
Preconditions	Effects	Cost
Dry(?tbr)	Painted(Table), ¬Dry(?tbr)	5.0

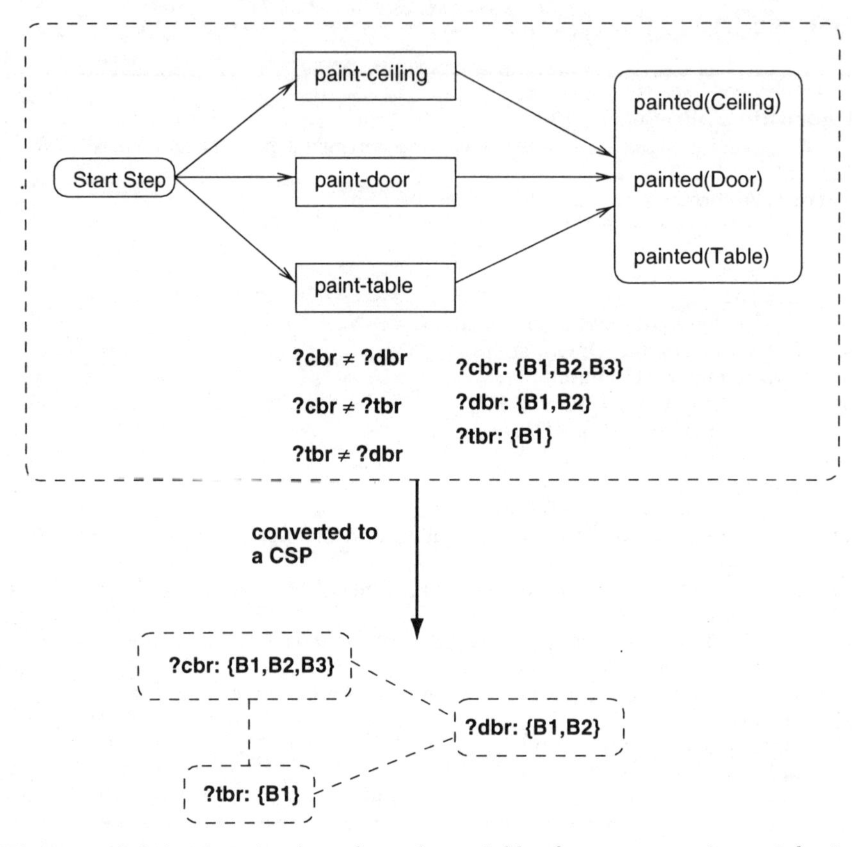

Fig. 5.4. A finite-domain plan where the variables form a constraint satisfaction problem

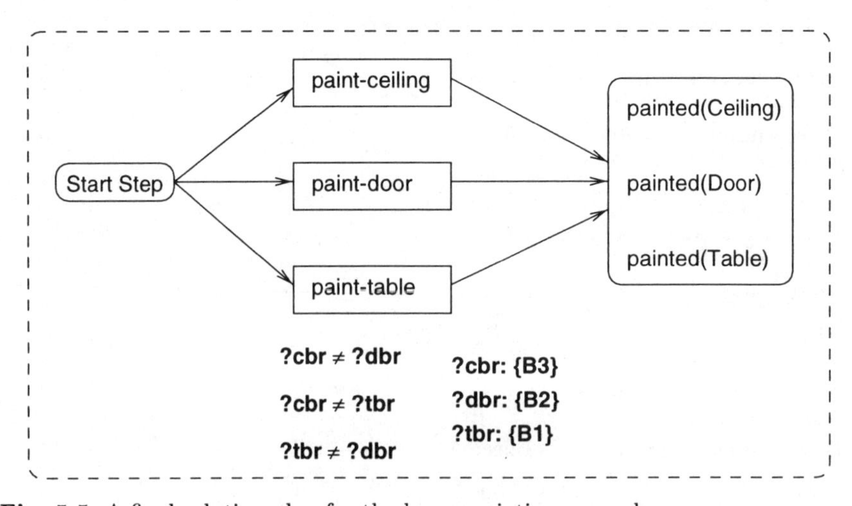

Fig. 5.5. A final solution plan for the house-painting example

this plan every precondition has an associated causal link, and Step 5 of FDPOP algorithm is encountered. At this point, a CSP solver processes a CSP in Figure 5.4. When examining a variable pair, (?tbr, ?dbr), arc-consistency found that the value B1 can be removed from the domain of the latter. Similarly the value B1 is removed from the domain of ?cbr. When the pair (?dbr, ?cbr) is processed, the value B2 is removed from the domain of ?cbr. The resultant singleton domains consititute a final solution to the CSP. And the corresponding solution plan is shown in Figure 5.5.

This example demonstrates several advantages of FDPOP. Suppose that during planning, we introduce a plan step with a variable $?x$, where $?x$ can be bound to n different constants, $a1, a2, \ldots, an$. If we next insert another step with a parameter $?y$ that can be instantiated to any of m constants $b1, b2, \ldots bm$, then the state space of a traditional partial-order planner might look like Figure 5.6a. In this search tree, the partial-order planner commits to one of $a1$ through an at one level, and then commits to one of $b1$ through bm at another level. In the worst case, there are tight constraints on their values which can only be satisfied by constants an and bm, and the entire search tree with a size of $m * n$ must be scanned.

In contrast to the traditional approach, the extended constraint-based planner does not commit to any specific constant during planning. The domain values for both $?x$ and $?y$ are simply recorded and reasoned about at a later time. Therefore the search tree looks like the one shown in Figure 5.6b.

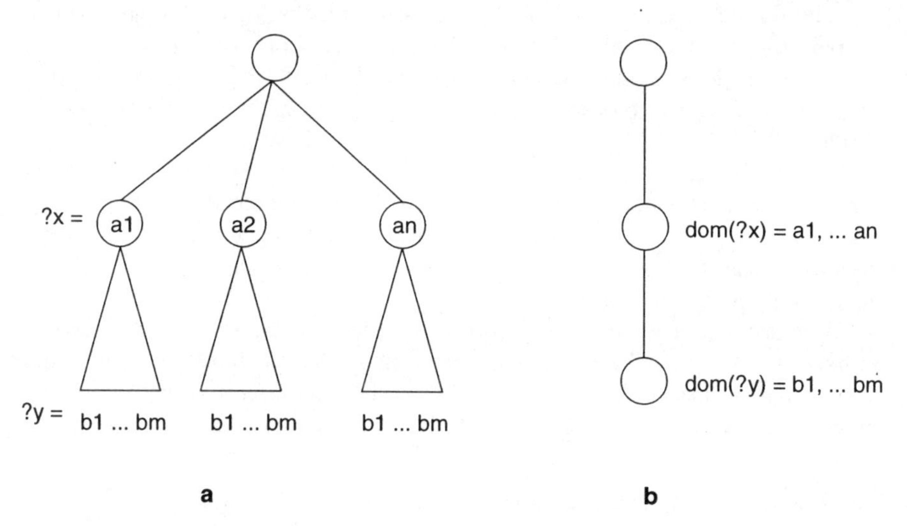

Fig. 5.6. Comparing TOPPOP and FDPOP

5.5 A Variable-Sized Blocks World Domain

In the above we have shown a representational method for reasoning about finite resources available to planning operators. An important remaining question is *how often* constraint satisfaction routines should be applied during planning. We have several options in implementing the constraint-checking part of step 15 in FDPOP. Here we could choose to verify whether a solution exists to the CSP in every plan expansion, every other expansion, or every n^{th} expansion. It is a well-known *utility question* to determine which option works best.

In this section, we do not attempt to address the utility question fully, but instead present an illustrative example and show how the utility issue for other planning domains could be similarly investigated.

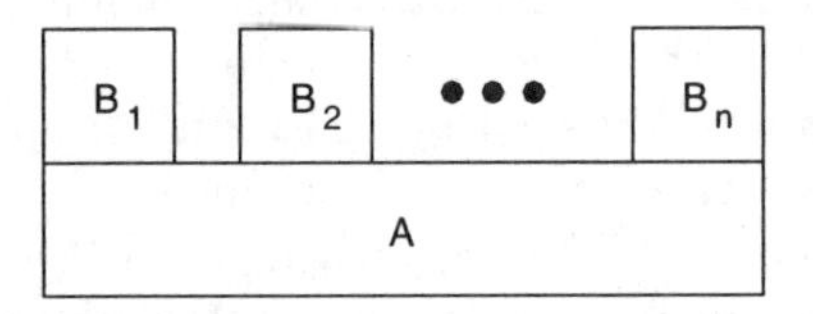

Fig. 5.7. The variable-sized blocks domain

The domain we look at is a variable-sized blocks world domain. In this domain there are a number of blocks $B_1, B_2, \ldots, B_n$ of unit size, and there is a particular block A that can support more than one other block. The task is to move all the B blocks on top of block A, using the operator shown in Table 5.4.

In this definition, L_i represents a location available on block A. The variable $?s$ could be bound to any location on A. By varying m, we can change where a *dead end* occurs in the search tree, as well as the percentage of correct and consistent solutions among the leaf nodes of the tree. We call the latter the density of solutions.

In this domain we ran an empirical test, where the number of B blocks is fixed at five, and the number of locations on top of block A varies from two to five. A dead end occurs very high up in the search tree if the number of locations is two; it occurs relatively deep in a search tree if the number of

Table 5.4. Operator definition with resources

`PutonA`$(?x, ?s)$
Var: $?x : \{B_1, B_2, \ldots, B_n\}$; $?s : \{L_1, L_2, \ldots, L_m\}$;

Precondition	Effect	Cost
$\text{Clear}(?x), \text{SpaceOnA}(?s)$	$\text{OnA}(?x), \neg\text{SpaceOnA}(?s)$	1

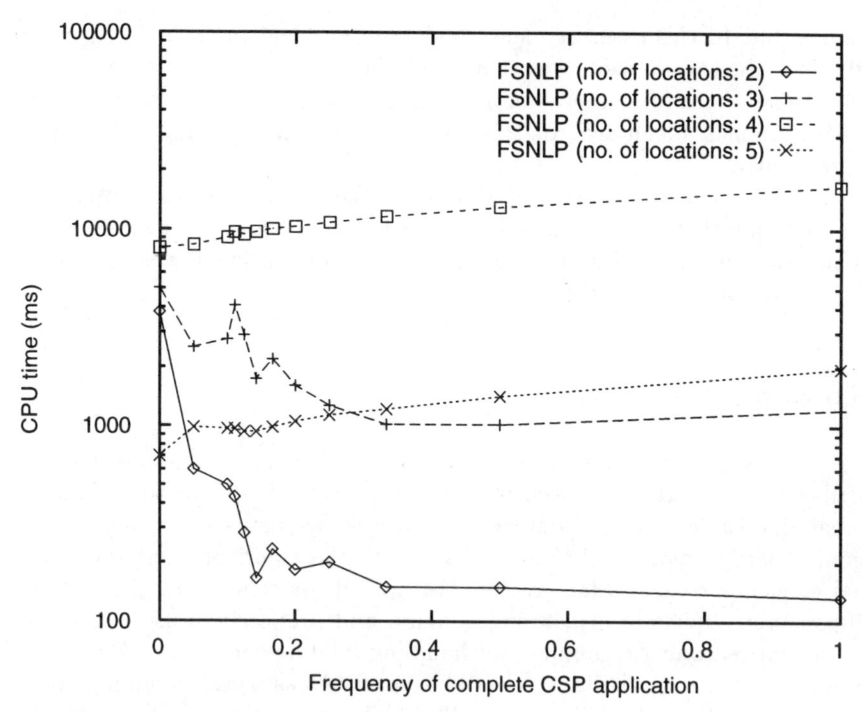

Fig. 5.8. Performance in variable-sized blocks domain

locations is four. When the number of locations is five, all five blocks can be put on block A at the same time. In this case no dead end occurs.

Our implementation of FDPOP is an extended SNLP algorithm [17], which is run on a SUN4/Sparc Station using Allegro Common Lisp. We call the resulant planner FSNLP.

Figure 5.8 depicts the empirical results, where the *frequency* of CSP applications refers to the inverse of the number of nodes, minus one, along a search path between any two successive CSP applications: the frequency is one if CSPSOLVE is applied to every node, two if every other node. If CSPSOLVE is not applied to any node at all, then the frequency is 0. The figure shows that when the dead end occurs early in the search tree, for example, when the number of locations is 2 (the solid line in the figure), the CPU time consumed is in inverse proportion to the frequency of CSP application. In the opposite case, when the number of locations is many (five), the situation is reversed: the CPU time used is in direct proportion to the frequency of solving the constraint network, implying that it is better to apply CSP checking infrequently.

The result makes sense intuitively. If the number of locations on block A is few, then the dead end occurs high up in the search tree, and therefore when CSPSOLVE is applied frequently these dead ends are detected as soon as possible. In this case, frequent application of CspSolve could save a lot of effort. In the opposite case, most of the tree paths in the search tree lead to solutions, and therefore applying CspSolve will not provide much computational advantage. In this case, it is better to apply CSP solving infrequently.

Lastly, it should be emphasized that the utility question is still very much an open question. Although we have shown that in this restricted domain it is possible to understand the impact of the problem fairly well, in general much more research is still needed.

5.6 Summary

As a case study in applying constraint satisfaction to planning, we have described how to extend a basic planning algorithm, Poplan, to include the capability for reasoning about finite resources. We can see that such extension, in fact, requires only minor changes to the algorithm, but the potential efficiency return is far greater. We argued for collective representation of constant objects in a planning domain, and proposed to use well-known constraint reasoning techniques for handling binding constraints. Finally, we provided empirical results in a variable-sized blocks world domain, demonstrating how the utility issue regarding CSP applications could be investigated.

In subsequent chapters, we will continue to use the basic planning domain representation, introduced in Chapter 2. This is mainly due to its simplicity. We should bear in mind, however, that these subsequent techniques in resource reasoning could in fact be extended in a conceptually simple manner.

5.7 Background

Reasoning about constraints is a major aspect of knowledge-based scheduling; a comprehensive coverage can be found in [50], [51] and in a collection of articles [52]. Stefik demonstrates that it is more advantageous to combine planning and resource reasoning [128]. The Molgen system functioned mainly in the domain of biological and molecular experiments, and as such it had certain limitations. Stefik's experiments, for example, demonstrated that a certain partial arc-consistency processing in reasoning about variable-binding constraints could lead to efficiency gain, but did not address resource reasoning in the context of a formal model of constraint satisfaction.

Constraint-directed reasoning is becoming a necessary component of real-world AI planning systems. Examples of successful combination of the techniques include Wilkins' Sipe [147], Lansky's Gemplan [88], and Ghallab and Laruelle's IxTeT [56]. Constraint reasoning in temporal reasoning has also been one of the research themes of Allen [7, 3]. Joslin and Pollack proposed a method for representing and reasoning about *all* planning decisions as constraints [69]. Finally, Kautz and Selman proposed a method for encoding actions as state axioms, and use them in a GSAT-based planning algorithm [77].

The work reported in this chapter is adapted from an AIPS94 paper by Yang and Chan [155]. More experimental results can be found in Chan's Master's thesis [26].

Part II

Problem Decomposition and Solution Combination

Overview

Profound and focused
and we will be strengthened,
Confused and divided
and the opponents are weakened.

Focused and we act as one
Weakened as they divide in ten.

(Sun Tzu, The Art of Strategy, *Chapter Six)*

In this part we provide a computational framework for practising **divide-and-conquer**. We first discuss how to decompose a domain automatically in Chapter 6, based on an analysis of the inherent interactions among the goals. The result of the decomposition process would be a partition of a set of goals into subsets, such that the goals in each subset could be solved concurrently. When they are combined, the solutions to the subsets might still have conflicts between each other. Thus, in Chapter 7 we develop a constraint-based method for resolving the conflicts efficiently. Moreover, in Chapter 8 we discuss how to take advantage of the overlapping capabilities between solution plans by merging them. Finally, Chapter 9 deals with the problem of how to select solution plans from among a pool of candidates.

6. Planning by Decomposition

The ability to decompose a complex problem into manageable sub-components is a necessity for many intelligent problem-solving activities. Presented with a complex planning problem, an intelligent agent following the *divide-and-conquer* methodology tries to decompose it into subproblems, so that each can be solved separately. The solutions are then reassembled for an overall solution. When performing the decomposition, careful attention is paid to the partitioning process so that clean interfaces with controlled interactions remain. A very important issue is to contain the complexity, and limit the number and variety of mechanisms needed to solve each subproblem. In this chapter we outline an approach to planning using decomposition and discuss its differences and similarities with other planning algorithms.

6.1 Decomposition Planning

6.1.1 Two Examples

A Household Example. We begin by considering the following household example. Suppose that on a weekend, we plan to repaint our house and do some grocery shopping. Our list of subgoals are the following:

Filled(Ceiling-Hole), Painted(Ceiling), Painted(Window),
Painted(Wall), Painted(Door), Have(Bread), Have(Fruit),
Have(Vegetables), Have(Milk), Have(Detergent), ...

To work on the ceiling and to complete the painting jobs, we need to wear a pair of gloves, to change to working clothes, to get a ladder, and to get the painting brushes and paint. To do grocery shopping we need to wear somewhat better clothes, to drive a car to the supermarket, and to get each item in turn. In the end, we have to pay for all these items.

A sensible agent is likely to partition these tasks into two classes, one for house painting and the other for shopping. When planning for this household problem, a plan can be separately developed for achieving the goals in each class, and, in the end, the two plans can be combined.

A Multi-table Blocks World Example. Next, we consider a scenario where there are 10 tables, $\text{Table}_i, i = 1, 2, \ldots, 10$. On each table i is a stack of blocks, A_i, B_i, etc. Our goal is to arrange the stacks on the tables from their initial configurations to their goal configurations. We have one restriction, though, requiring that no block can leave its own table.

In this domain, our intuition tells us that it is better to develop plans for transforming block-configurations on each table, in a concurrent manner. This decision is made despite the contention among the ten tasks for the robot hand as a resource.

6.1.2 Global Planning-Domain Decomposition

In both problems above, we decided to decompose a large set of goals into several subsets and work on each subset separately from the rest. The decomposed goals entail a partition of the planning domain into disjoint subsets, where each subset has its own operators and initial state. A global-decomposition planner develops a plan for each subset of goals concurrently with others. A subsequent step resolves conflicts among the separately developed plans and merges these plans.

More precisely, we advocate a global-decomposition planning strategy. Let a planning problem be $\langle init, goal, \mathcal{O} \rangle$, where $init$ is a set of initial state literals, $goal$ are the goal literals, and $\mathcal{O}$ is the set of domain operators. Assume that the problem is decomposed into m parts, such that for each part $\langle init_i, goal_i, \mathcal{O}_i \rangle$, the initial state $init_i$, goal state $goal_i$, and operators $\mathcal{O}_i$ are subsets of $init$, $goal$, and $\mathcal{O}$, respectively. Furthermore, assume that the union of initial state $init_i$, goal state $goal_i$, and operators $\mathcal{O}_i$, for $i = 1, 2 \ldots m$, give rise to the original sets $init$, $goal$, and $\mathcal{O}$, respectively. The algorithm for global-decomposition planning is shown in Table 6.1.

There are five critical components in this algorithm. The first component is a method for decomposing a domain into several subparts. This is done by the algorithm GDECOMP. We will present an analysis of the benefits of decomposition in the next section. Through this analysis we provide some guidelines on good domain decomposition which can lead to improved planning efficiency. The second component in this algorithm is a method for finding solutions for each decomposed sub-domain. To implement this component, we could employ any of the basic algorithms discussed in Chapter 2; a criterion on which algorithm to select is also discussed in the next section. The last three components are for selecting a solution for each sub-domain, resolving conflicts among the sub-solutions and merging plans by removing redundant parts in the combined solutions. These three components are discussed in detail in Chapters 9, 7, and 8, respectively.

Table 6.1. The DECOMPLAN algorithm

Input: A planning domain and a planning problem (*init*, *goal*).
Output: A solution plan Π for solving the planning problem.

Algorithm DECOMPLAN($\mathcal{O}$, *init*, *goal*)
1. apply the decompositon algorithm GDECOMP() , obtaining m subproblems $\{\mathsf{prob}_i = \langle init_i, goal_i, \mathcal{O}_i \rangle, i = 1, 2, \ldots m\}$;
2. For each problem prob_i, find a set of solutions $\{\Pi_{ij}, j = 1, 2, \ldots, k\}$;
3. Select one solution plan Π_{ij} for each subproblem; let the selected solutions be Solution;
4. Resolve conflicts among the plans in Solution;
5. Merge the plans in Solution;
6. If steps 3 and 4 are successful, return the resultant solution plan.
7. If every possible combination of solutions has been considered, exit with failure. Otherwise, go to step 2.

6.2 Efficiency Benefits of Decomposition

One of the early analyses was provided by Korf [83], who discussed computational complexity issues related to decomposing a compound goal into subgoals. His analysis is based on several strong assumptions, and is implicitly based on forward-chaining, total-order planners. Below, we review the key points in his analysis and present our extensions. In the subsequent analysis, we use `No-Decomp` to denote a generic planner which solves a conjunctive goal without using decomposition, and `Use-Decomp` for a planner which uses decomposition.

6.2.1 Forward-Chaining, Total-Order Planners

Korf's Analysis. First, suppose that to solve each subgoal g_i separately, the branching factor of a forward-chaining, total-order planner is F_i. That is, in the process of achieving g_i, for any state there are F_i applicable operators. Suppose also that the optimal solution length for a subgoal g_i is N_i. Then when all the goals are solved together using a forward-chaining planner, at each state all applicable operators must be used to generate successors. When all subgoals are completely independent, the worst-case time complexity can be derived from equation (3.2) in Section 3.2:

$$\text{Time}(\texttt{No-Decomp}) = O((\sum_{i=1}^{m} F_i)^{\sum_{i=1}^{m} N_i} * T_f) \quad (6.1)$$

$$= O((m * F)^{\sum_{i=1}^{m} N_i}) \quad (6.2)$$

In the above formula, F is the maximum of F_i, for $i = 1, 2, \ldots, m$. The T_f factor denotes the *constant* time spent in each node expansion.

Now suppose that, as Korf proposed, we decompose the goals into m subgoals, solve each subgoal in turn, and then combine all solutions by concatenation. Since all m goals are completely independent of each other, the worst-case time complexity is

$$\text{Time}(\texttt{Use-Decomp}) = O(\sum_{i=1}^{m} F_i{}^{N_i} * T_f) \quad (6.3)$$

$$= O(m * F^N) \quad (6.4)$$

where N is the maximal optimal solution length for the subgoals.

Based on the above formulas, Korf's conclusion is therefore that in the case of complete independence, and when forward-chaining planners are used, decomposition cuts down both the branching factor and the depth of search, leading to an exponential amount of savings in planning time.

Interacting Goals. We now extend Korf's analysis to interacting subgoals. Now, instead of requiring that the operators for each goal g_i be completely independent of the operators for any other goal g_j, suppose that we allow a certain number of interactions. In particular, we allow

1. an operator relevant for achieving one goal to *delete* the precondition of an operator relevant for another goal. This is called a *deleted-condition interaction.* Deleted-condition interactions are caused by the presence of negative threats in the plans. They are also called *conflicts.*
2. an operator relevant for achieving one goal to *add* the precondition of another operator. This is called an *operator-overlapping interaction.*

As an example, consider two plans for painting the ceiling and painting the ladder, shown in Figure 6.1. The operators `get-ladder`, one from each plan, can be combined into a single one. This is an example of the operator-overlapping interactions. The `paint-ladder` operator interacts with the `paint-ceiling` operator through a deleted-condition interaction; painting the ladder might leave the ladder with wet paint, making it unusable for painting the ceiling later.

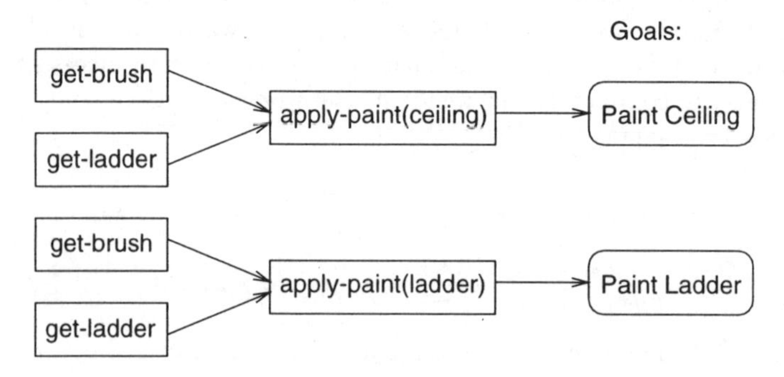

Fig. 6.1. Interactions in the household domain

When these two interactions are both present, the analysis becomes more complex. We now not only have to consider the cost for solving the individual goals, but also the cost for combining the solutions to form a global solution.

A. No Decomposition. We analyze the search behavior of a forward-chaining, total-order planner when all goals are solved together. Due to the *operator-overlapping interactions*, some operators relevant for achieving one goal may themselves be part of operators relevant for achieving another goal. Thus, when all goals are solved together, the branching factor of search is somewhat smaller than the sum of all individual branching factors F_i.

As for the search depth, both the operator-overlapping interactions and the deleted-condition interactions will contribute to it in various ways.

First, due to operator-overlapping interactions, some operators may serve to establish more than one subgoal. Therefore, the effect of operator-overlapping interactions is that a certain number of operators N_o can be subtracted from the sum $\sum_{i=1}^{m} N_i$ when computing the total solution length.

Second, due to deleted-condition interactions, some operators for achieving one subgoal in a solution may delete a precondition of another operator for achieving a different subgoal. To patch up the deleted preconditions, extra operators may have to be inserted. Therefore, the effect of deleted-condition interactions on total solution lengths is that a number N_d of operators have to be *added* to the sum $\sum_{i=1}^{m} N_i$ when computing the total solution length.

Putting the above factors together, when solving all subgoals at the same time, the worst-time search complexity is

$$\text{Time}(\texttt{No-Decomp}) = O\left((\sum_{i=1}^{m} F_i - F_o)^{(\sum_{i=1}^{m} N_i + N_o - N_d)}\right) \tag{6.5}$$

B. Decomposition. With decomposition, the planning process is completed in three stages.

Stage 1. The set of all goals is decomposed into multiple subsets $g_i, i = 1, \ldots m$;
Stage 2. For each goal set g_i, a solution is found;
Stage 3. The solutions to the m goal sets are combined into a global solution.

It is also possible that at stage 3 we meet a dead end and therefore the solutions cannot be combined into one that achieves all goals. In that case, alternative solutions must be planned for in stage 2.

We assume that the computational complexity for performing the decomposition is Time(GDECOMP), where GDECOMP is a goal-directed decomposition algorithm. The complexity for the second stage is the same as the case when there is complete independence; to solve each subgoal using a forward-chaining, total-order planner, the time complexity is $O(F^{N_i})$. Here we concentrate on the complexity for the third stage, the *solution-combination* stage.

Let I_d be the total number of deleted-condition interactions among the actions for achieving the goals. To resolve these interactions the worst-case time complexity is $O(r^{I_d})$, where r is the number of ways to resolve each conflict. We assume that $r > 1$.

Related to the deleted-condition interactions, there are also I_o operator-overlapping interactions. For each possible operator overlap we have to decide whether to "merge" the two operators into one. For example, in the household domain, it is possible to get a ladder only once while satisfying a precondition for both painting the ceiling and painting the wall. In the worst case, the time complexity for taking care of all operator-overlapping interactions is $O(2^{I_o})$, which is also exponential in I_o.

Between deleted-condition interactions and operator-overlapping interactions, there is also a certain amount of interaction. One way of resolving a deleted-condition interaction may turn out to make a merging operation impossible. Therefore, when both types of interaction are present, the worst-case time complexity for resolving both is $O(r^{(I_d+I_o)})$.

Finally, in stage 3 we may fail to find a consistent solution to all goals. In that case, a certain amount of backtracking to stage 2 must occur. Let k be the number of plans for a goal g_i. A total k^m combinations might have to be considered in the worst case before finding a consistent global solution. Thus, the time complexity for solving the planning problem with decomposition, using a *forward-chaining, total-order planner*, is

$$\begin{aligned} \text{Time}(\texttt{Use-Decomp}) \quad = \quad & O(\text{Time}(\text{GDECOMP}) + \\ & (m * F^N + r^{(I_d+I_o)}) * k^m) \end{aligned} \tag{6.6}$$

C. Summary: Interacting Goals. When the goals are interacting, and when a forward-chaining, total-order planner is used, decomposition no longer has definite advantages. For decomposition to win, the number of operator-overlapping interactions and deleted-condition interactions must be small. Low operator-overlapping interactions imply a small I_o value in equation (6.6) and small N_o in equation (6.5). Similarly, low deleted-condition interactions imply a small I_d value in equation (6.6). Low interactions also make it more likely that an early selection of solutions could be successfully combined. This latter fact leads to a small value in the k^m factor in equation (6.6).

To sum up, a "good" planning-domain decomposition should ensure that the number of operator-overlapping interactions and deleted-condition interactions between the decomposed domains are small. In this case a decomposition planner could still reduce the search time by decreasing both the branching factor of search and the depth of search.

6.2.2 Backward-Chaining, Partial-Order Planners

Korf's analysis did not take into account the possibility of a backward search. The analytical formula for a backward-chaining, partial-order planner is different because, unlike forward-search planners, the branching factors are determined by the number of operators that can achieve a condition, rather than the number of operators that are applicable to a state. The search depth is also different. The following analysis will be based on the analytical formula derived in Chapter 3.

A. No Decomposition. We start by assuming that the goals are interacting. Suppose that for a goal g_i, the branching factors are $B_{new,i}$ and $B_{old,i}$, representing the maximal number of *new* operators and *existing* plan steps that could achieve a precondition in a plan, respectively. Suppose also that to achieve a goal g_i, the optimal number of steps (discounting the start and final steps) is N_i. As in the forward-chaining case, we must account for the deleted-condition interactions and operator-overlapping interactions among the plans. The former adds a N_d factor to the total solution length, while the latter subtracts a N_o factor. Let L be the sum of individual solution sizes: $L = \sum_{i=1}^{m} N_i$. Recall that for backward-chaining, partial-order planning, $B_b = [(B_{new} + B_{old}) * r^t]^P$, where t is the maximal number of threats in each plan, r is the number of ways to resolve each conflict, and P is the maximal number of preconditions for each plan step. Let B_b be the maximum such value among all goals.

The time complexity can therefore be easily derived from equation (3.5) in Chapter 3:

$$\text{Time}(\texttt{No-Decomp}) = O\left({B_b}^{(L-N_o+N_d)}\right) \tag{6.7}$$

B. Decomposition. With decomposition, partial-order planners follow exactly the same three stages as in the forward-chaining case. Suppose that, as before, each goal has k alternative solution plans. Let N be the maximal number of plan steps (except the start and the finish steps) in a solution plan for solving an individual subgoal g_i. Then for *backward-chaining, partial-order planners*, using decomposition has a time complexity of

$$\begin{aligned}\text{Time}(\texttt{Use-Decomp}) &= O(\text{Time}(\text{GDECOMP}) + \\ &\qquad [m * {B_b}^N + r^{(I_d+I_o)}] * k^m)\end{aligned} \tag{6.8}$$

C. Summary: Interacting Goals. When the goals are interacting, and when a backward-chaining, partial-order planner is used, we reach a similar, but not identical conclusion as in the total-order case. For decomposition to win, the number of operator-overlapping interactions and deleted-condition interactions must still be small. Low operator-overlapping interactions imply a small I_o value in equation (6.8) and a small N_o in equation (6.7). Similarly, low deleted-condition interactions imply a small I_d value in equation (6.8). Low deleted-condition interactions also increase the chance for an arbitrary selection of solutions to be successfully combined, thus reducing the k^m factor in equation (6.8).

However, we see that unlike the total-order planning case, planning both using and not using decomposition may have the same effective branching factors. This happens when the operator sets for the decomposed planning problems consist of the identical operators. However, *a potential efficiency benefit for decomposition is likely derived from a reduction of the search depth.*

6.2.3 Criteria for Good Decomposition

A good decomposition should enable a planner to reduce the planning time. Based on the analysis for both total-order and partial-order planners, we see that when a "good" planning-domain decomposition is used, it is potentially possible for a planner to gain an exponential amount of efficiency by a reduction of the search depth. For forward-chaining, total-order planners, an additional advantage is that the branching factor could be reduced. From both analyses, we see that a criterion for a good decomposition is that both operator-overlapping interactions and deleted-condition interactions be small.

Note that for an arbitrarily chosen decomposition there is no guarantee that planning efficiency will be increased. Factors such as the number of operator-overlapping interactions and deleted-condition interactions play a central role in determining whether a decomposition is good.

6.3 Goal-Directed Decomposition: The GDECOMP Algorithm

Goal-directed decomposition refers to the task of decomposing a large set of goals into smaller subsets, such that each subset of goals is planned for together, and that different subsets are planned for separately. We start designing the decomposition algorithm by re-visiting the intuition in the household domain.

Suppose that the tasks for painting the house and furniture, and for doing grocery shopping, are represented as goals. One can easily imagine many such goals, the number of which can be quite large. Our intuition tells us that these goals could be roughly split into two groups, one for painting only and the other for shopping. The reason behind this grouping is that the actions *relevant* for the painting goals are highly related with each other, while they are less related to the actions relevant for shopping. Our conclusion here is that analysis of the potential interactions among the goals can provide sensible goal-directed decompositions.

We try to capture this intuition in our decomposition algorithm GDECOMP (see Table 6.2). Below, we present this algorithm in a succession of steps. In the process, we assume that the definitions for all operators are given along with all constants in the domain. We assume that all operators are fully instantiated; the extension of the subsequent discussion to un-instantiated operators is straightforward. We also assume that the interactions among the goals are of a binary nature; only pairs of goals can interact. Finally, we assume that an initial state description and goals $g_i, i = 1, 2 \ldots n$ are given.

Step 1. Compute Relevant Operators. For a goal g_i, only some operators are "relevant" to achieving it, either directly or by asserting the precondition of an operator which in turn is relevant to the goal. We capture this recursive definition for a **relevant set** of goal g_i, as follows:

Table 6.2. The GDECOMP algorithm

Input: A set of operators $\mathcal{O}$, initial state *init* and goal $g = \{g_1, g_2, \ldots, g_n\}$.
Output: A set of subsets, $\mathcal{G}_i$, each consisting of the goals to be planned for together.

Algorithm GDECOMP(O, *init*, g)
1. For each goal g_i in goal set g, compute a set of relevant operators.
2. **for** every pair of goals g_i and g_j, $i \neq j$,
   Estimate operator-overlapping interactions between g_i and g_j.
3. **for** every pair of goals g_i and g_j, $i \neq j$,
   Estimate deleted-condition interactions between g_i and g_j.
4. Fix a problem-solving model (forward or backward), and construct a goal graph.
5. Partition the goal graph, producing sets $\mathcal{G}_i, i = 1, 2 \ldots, m$.
6. **return**($\{\mathcal{G}_i, i = 1, 2, \ldots m\}$).

Definition 6.3.1 (Relevant Operators) *Let g be a goal and $\alpha \in \mathcal{O}$ be an operator in a domain. The operator α is relevant to goal g, or $\alpha \in Rel(g)$, if and only if*

1. *α directly achieves the goal: $g \in \text{Eff}(\alpha)$; or*
2. *α achieves a precondition of some operator $\beta \in Rel(g)$.*

Steps 2 and 3. Estimate the Interactions. We consider each pair of goals g_1 and g_2 and estimate the number of operator-overlapping and deleted-condition interactions between them. One way to obtain such an estimate is using a library of past plans. By counting the average number of such interactions between previous solutions for g_1 and g_2, a crude estimate could be obtained for $I_o(g_1, g_2)$. Ultimately a machine learning method should be used for computing these estimates. Based on previously generated plans for each goal, a prediction can be made for a future problem given that the planning-problem distribution stays the same. In subsequent discussion we assume that the estimates are known already.

Step 4. Fixing a Planning Model: Forward or Backward. We must now decide on whether to use a forward-chaining or a backward-chaining method for solving the goals. A different method has a different computational complexity formula (See Chapter 3). For purposes of exposition, here we assume that a forward, total-order planner is used.

We make the following assumptions:

- let B_i be the branching factor of search for a subset of goals G_i;
- let D_i be the optimal solution length for g_i;
- let B be the branching factor, and D the depth, of not using the decomposition in planning;

- let $\mathcal{I}(G_i, G_j)$ be the sum of all interactions between the members of the two goal sets; $\mathcal{I}(g_i, g_j) = I_d(g_i, g_j) + I_d(g_i, g_j) + I_o(g_i, g_j)$;
- let r be the maximal number of ways to resolve each conflict.

We require that

$$\sum_{i=1}^{m} B_i^{D_i} + r^{\mathcal{I}} < B^D, \text{ where } \mathcal{I} = \sum_{i,j=1,1}^{m,m} \{\mathcal{I}(G_i, G_j), i \neq j\} \tag{6.9}$$

Under these conditions, it can be shown that a goal-directed decomposition planner has a lower worst-case complexity when (a) the first choice of plans, one for each goal set, can be consistently combined ($k = 1$ in equation (6.8)) and (b) the computation for executing the GDECOMP algorithm is negligible. Again, one should eventually obtain the above factors through machine learning.

Step 4. Partition the Goal Graph. In this final step we use a greedy algorithm for performing the decomposition. This algorithm is based on a graph analysis using the estimated parameters.

We first construct a graph of goals, where every node is a singleton set containing the goal. Between every two nodes is a weighted edge, where the weight represents the number of interactions between the pair; the weight between any two nodes g_1 and g_2 is $\mathcal{I}(\{g_1\}, \{g_2\}) = I_d(g_1, g_2) + I_d(g_2, g_1) + I_o(g_1, g_2)$.

If equation (6.9) is not yet satisfied, we wish to find a pair of nodes (n_1, n_2) such that if they are combined into a single node, the left hand side of equation (6.9) decreases. To decide which nodes to combine next, we assume that nodes n_1 and n_2 are chosen. Let the interaction measure between n_1 and n_2 be $\mathcal{I}(n_1, n_2)$, and the sum of the interaction measures among all the nodes be $\mathcal{I}$. We would like to reduce the left hand side of equation (6.9) by a maximal number. Before the nodes n_1 and n_2 are combined, the size of the search tree expanded by a decomposition planner is $\sum_{i=1}^{m} B_i^{D_i} + r^{\mathcal{I}}$. After they are combined, the size becomes

$$\sum_{i=3}^{m} B_i^{D_i} + B_{12}^{D_1+D_2} + r^{(\mathcal{I}-\mathcal{I}(n_1,n_2))}$$

where $B_{12} = (B_1 + B_2)$ for forward, total-order planners.

To select n_1 and n_2, we require that the difference between the quantities is positive and maximal. This difference can be expressed by

$$(B_1^{D_1} + B_2^{D_2} + r^{\mathcal{I}}) - (B_{12}^{D_1+D_2} + r^{(\mathcal{I}-\mathcal{I}(n_1,n_2))}) \tag{6.10}$$

When all goals have the same branching factor and depth, this selection process is reduced to finding a pair of goals such that the number of interactions between them is maximal.

After n_1 and n_2 are combined to form a larger node $n_{12} = n_1 \cup n_2$, the weights in the resulting graph should be recomputed to reflect the change. For all other nodes n_3, the new weight between n_{12} and n_3 is the sum of the weights between n_1 and n_3, and n_2 and n_3. The process repeats until equation (6.9) is satisfied. The resulting graph contains the final decomposition of the goals.

6.4 Other Benefits of Decomposition

We have argued about good decomposition from an efficiency point of view. In this section, we briefly mention some of the other benefits of goal decomposition.

Concurrency in Plan Generation. By decomposing a problem domain into several parts, we could assign a planner to each part to generate a solution. The interactions among the parts could be represented as constraints among them, enabling more effective communications among planning activities [88].

Identifying Reusable Plan Parts. Suppose that an agent decides to reuse an existing plan. Suppose also that only a few goals in the current situation are different from the previous one. By having the domain decomposed we could quickly identify the parts of the plan affected by the change. In this case, only a limited amount of effort need be spent on finding sub-plans for the affected goals.

Multi-agent Plan Execution. A solution plan to a decomposed planning problem could be easily assigned to multiple agents to execute. The limited interaction among the agents ensures that there is a large degree of independence in each agent's action. In addition, confronting the possible interactions among the agents also makes it possible to formulate communication channels between them.

Balancing Achievable Goals. It is sometimes not economical or even possible to achieve all goals presented to a planner. In that case it is beneficial to identify a subset of goals that can be achieved. This decision can be made naturally when the goals are already decomposed into groups, so that between any two groups the interactions are limited. When many goals are to be achieved, the decision on which goals to keep and which to drop can be made more informative when the number of interactions among the goals is pre-estimated.

6.5 Alternative Approaches to Decomposition Planning

Divide-and-conquer has been a main theme of planning research. Although the main idea has always been to break apart a complex problem into simpler parts, the approaches vary drastically. Below, we consider some major thrusts in planning that takes decomposition into account.

Decomposition Along a Time Line

The first work we consider is planning with abstraction. With this type of problem solving, a domain is broken down into several levels on a hierarchy. The most critical part of a planning problem is represented at the highest level, and a solution obtained at the lowest or concrete level is a solution to the original problem.

Abstraction planning starts by finding a solution plan at the highest level and gradually refining the solution to each lower level. An early example of an abstract planner is ABSTRIPS [113].

During planning, an abstract solution defines a collection of subproblem descriptions that are separated by time. The intuition is that most of these subproblems will not interact, allowing them to be solved more or less independently. The abstraction paradigm was later extended in Knoblock's ALPINE [80], and Bacchus and Yang's HIPOINT [13], which will be discussed at greater length in Chapter 11.

Despite the similarity in the general framework, major differences remain. The first difference is the manner in which a final solution plan is derived. In abstraction planning, a complete abstract plan is first found at the topmost level. This abstract plan ignores some less important preconditions and goals. Once found, it is taken down to a lower level where these other preconditions and goals are considered and achieved. This top-down process, called refinement, continues until a solution plan is found at the lowest level.

During the refinement process, the abstract plan defines a decomposition of the original problem *along a time line*, as shown in Figure 6.2. In this figure an abstract plan $A \mapsto B \mapsto C$ is being refined at a lower level of abstraction. At the lower level, the abstract plan acts as a skeleton along the time line from the initial state to the goal state, such that between every pair of abstract plan steps is a new subproblem to be planned for.

Decomposition planning, on the other hand, functions in quite a different manner. As shown in Figure 6.3 the planning process occurs *concurrently*, instead of sequentially as dictated by an abstraction hierarchy. The solutions are formed in parallel, and later combined by interleaving operators and imposing additional constraints. Therefore, when each sub-plan is formed, there

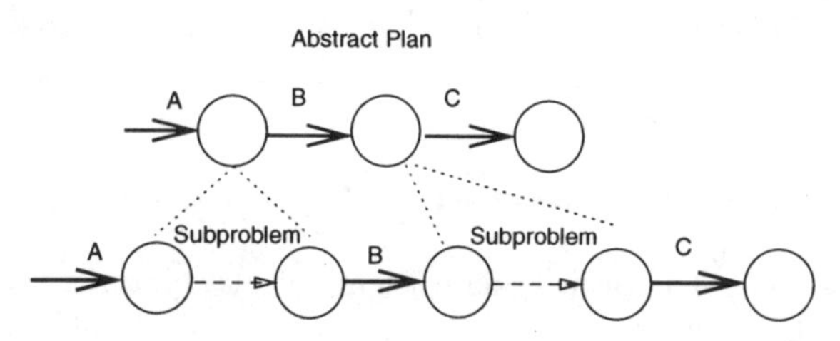

Fig. 6.2. Abstraction planning

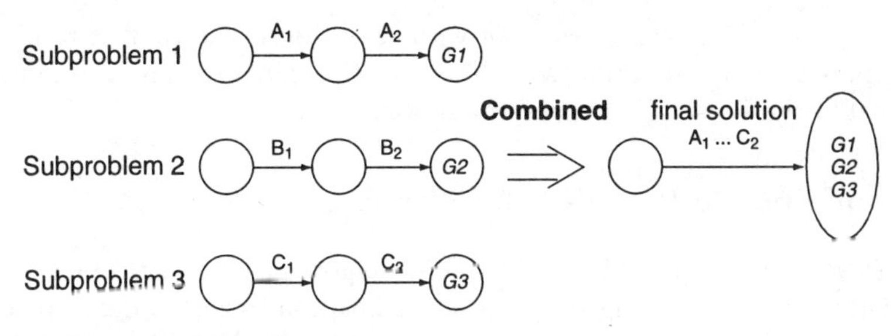

Fig. 6.3. Decomposition planning

is no elimination of preconditions as done in abstraction. The decomposition here is *not* along the time line, but instead, across the time line.

A second difference between abstraction and problem-domain decomposition, pointed out by Lansky[87], is that in obtaining the abstraction hierarchy used by ALPINE and HIPOINT, a requirement is enforced such that operators at a lower abstraction level cannot interfere with an abstract plan via interactions. This requirement is relaxed in decomposition planning, where we allow a non-empty set of interactions to occur between different groups of goals. These remaining interactions will be taken care of at a later stage, when the solution plans are combined.

Decomposition in STRIPS Planning

STRIPS is an early planning system designed to solve conjunctive goals [46]. Given a set of goals to achieve, STRIPS decomposes them into a sequence of subgoals. It then selects the first subgoal in the sequence, and finds an operator to achieve it. The operator is likely to introduce new subgoals; those which are its preconditions. These subgoals are decomposed in the same manner. When a complete solution is found for a given subgoal, STRIPS executes that solution plan from the current initial state to obtain a new state in which the subgoal becomes achieved.

STRIPS can be considered to implement a naive decomposition planning algorithm; its subgoals are solved one at a time and solutions to the subgoals appended in a linear sequence. However, when solving one subgoal, STRIPS does not completely separate it from the other subgoals. The effects of solving a subgoal is always felt by all the other subgoals. Nor does it decompose the initial state into subsets. The solutions to subgoals are only appended to each other.

In contrast, the style of decomposition planning that we consider here separates the goals into subgoals as a result of analyzing the number of interactions among the goals. Thus, instead of always resulting in a decomposition where each subproblem contains only one conjunct, in decomposition planning each subproblem can be defined by multiplesubgoals that are more closely related with each other than with the others. The conflict-resolution and solution-combination steps in decomposition planning are also more sophisticated than a simple append operation.

Decomposition in Partial-Order Planning

Like STRIPS, most partial-order planning algorithms have taken a rather simplistic view of planning-problem decomposition, where a decomposition is based on individual conjuncts in the definition of goals and subgoals given in the input. Actions are added to a plan to achieve subgoals one at a time. When searching for a solution plan, the entire plan is still considered together for threat removal and step addition; the effect of achieving one goal has an immediate impact on the entire plan.

A consequence of the partial-order planning approach is that no concurrent problem-solving is done; the effect of any operator in a plan reaches across the entire plan. The problem-solving efficiency is gained only through the inherent partial-order and partial-binding representation of plans, but not from a reduction of the problem itself.

A related effort in partial-order planning is the work on classifying problem-solvers by Barrett and Weld [17]. As a result of this analysis, partial-order planning is shown to be most effective in solving several types of interactions among goals and subgoals. An example of such situations is when the goals must be interleavingly achieved. However, the emphasis in this analysis is not on decomposing a domain into concurrent parts, but rather on demonstrating that for a subclass of conjunctive goals partial-order planners are superior to some versions of backward-chaining, total-order planners. The work by Barrett and Weld can be seen as defining a set of detailed properties of subgoals as a function of planning algorithms. In contrast, our analysis presented in this chapter can be seen as methods for automatically ascertaining some of these properties.

Decomposition by Regions

Closely related to our decomposition-planning algorithm are the GEMPLAN and COLLAGE systems designed by Lansky et al. [88, 89]. This approach advocates an action-based planning framework by generating plans concurrently based on regions of activity. The regions, which are either provided by the user or computed using a set of action constraints, function as a decomposition of the problem domain. The planning processes in different regions communicate with each other via user-defined *constraints*. Lansky refers to this style of concurrent planning by local regions as *localization*.

GEMPLAN[88] and COLLAGE[89] do not generate a decomposition of the problem domain into regions. Generation is handled by another system LOC

[87], developed more recently. LOC takes as input a set of actions and constraints in a planning domain and produces a hierarchical organization of regions. It accomplishes this task by invoking a procedure to determine whether two sets of actions are closely related; they are closely related when they are bound by the same constraint. For example, the actions `paint-wall` and `repair-wall` might be under the same constraint: `before(repair-wall, paint-wall)`. In this case, the two actions belong to the same local region.

In this manner, LOC determines a hierarchy of regions that forms a parent-child relationship. Planning can occur concurrently in different local regions, which are coordinated by inter-regional constraints.

The GDECOMP algorithm is similar to GEMPLAN in spirit; they both implement planning as an instance of the divide-and-conquer methodology.

Decomposition for Distributed Plan-Execution

In distributed AI, a more recently developed group of planners are exclusively aimed at solving a complex problem by working on individual parts by multiple agents in a distributed way. Most of them, however, depend on the users to provide a decomposition *before* the algorithms can be used. One theme of work in distributed planning is to assign agents to tasks in a more or less optimal way. A characteristic example is the DMVT planner [41], which decomposes the agents' environment by ranges of camera angles, assuming that corresponding to each sensor a dedicated agent exists. The *Contract Net* technique [125] decomposes a problem domain for a given set of agents, using certain negotiation and bidding techniques. The aim in that work is to develop a common protocol to facilitate agent negotiations, rather than to decompose a domain using a centralized algorithm as we propose.

6.6 Summary

Divide-and-conquer is one of the problem-solving skills that an intelligent agent should possess. One instance of divide-and-conquer is to partition the goal set into a number of groups, such that the inter-group interactions are controlled.

In this chapter we have analyzed the benefits of problem decomposition. This analysis confirmed our intuition that decomposition can lead to improved problem-solving performance. However, we have also shown how decomposition can hurt problem-solving if not used properly. It is the task of an intelligent decomposition method to make sure that the resultant decomposition is of good quality. We presented a criterion for judging good quality in terms of a complexity analysis. In addition, we discussed various approaches to decomposition and possible future extensions.

6.7 Background

An early system that performed problem decomposition was STRIPS [46]. To solve a set of conjunctive goal literals, STRIPS takes every single goal literal and every subsequent precondition literal as a separate subgoal in the decomposed goal sets. The tradition was carried on to other types of planning method, including Waldinger's planner [142], Nilsson's DCOMP (see Chapter 8 of [102]), and many partial-order planners (e.g., TWEAK [28] and SNLP [17]). The ALPINE system is described in [80].

7. Global Conflict Resolution

In decomposition planning, a complex planning problem is split into several parts. Solutions to each part are generated and selected. When these disjoint solutions are combined to form a global solution to the original planning problem, conflicts between different sub-solutions will become explicit. These conflicts occur at this stage due to the assumptions made when solutions are derived for each subproblem. An example of such assumptions is one on resources; when generating solutions for a subproblem, one might assume that a certain resource is always available when in fact a solution for a separate subproblem also claims the resource. Because many conflicts may exist when the solutions are combined, it is necessary to design an intelligent conflict resolution method. This chapter presents a method for dealing with this problem.

7.1 Global Conflict Resolution — An Overview

Conflicts rarely exist alone. Therefore, conflict-resolution decisions should not be made in a local manner. To see this, consider a household example.

An Example

Suppose that one wants to paint a ceiling as well as a ladder. A proposed plan may consist of two parts, one for painting the ceiling and the other for painting the ladder, such that no ordering constraint is imposed between the two parts. If the robot hand can only hold one brush at a time, then a resource conflict occurs between the part of the plan that paints the ceiling and the part of the plan that paints the ladder, because of their competition for the robot hand. Similarly, if the wet paint from painting the ladder precludes one from climbing up, then another conflict occurs because performing the former negates a precondition of the latter, which requires that the ladder be dry.

In the above example, a conflict is caused by a single plan step which can potentially delete a precondition of another step. The resource conflict can be resolved by painting the ceiling either before or after painting the ladder. And the wet-paint-on-ladder conflict is resolved by painting the ceiling first.

A successful resolution of both conflicts involves the recognition of several constraints on the order of plan steps and variable bindings. A consistent solution for resolving both conflicts is to paint the ceiling first.

Overview of the Chapter

In this chapter, we present a conflict resolution algorithm based on constraint satisfaction. This requires a formal representation of the individual conflicts and their resolution methods, an analysis of the relations among different conflicts, as well as the design of reasoning techniques that can facilitate global conflict resolution. We present a formalization of conflicts and their inter-relations, using an extended framework for solving constraint satisfaction problems, or CSPs (for an introduction to CSP, see Chapter 4). The formalization makes it possible to apply many existing techniques from the CSP area to aid efficient conflict resolution in planning. An additional feature of global conflict resolution is its extension to traditional methods for solving CSPs by employing a new constraint known as *subsumption relations.* These relations help remove a large portion of redundant constraints from the search space. Finally, we explore the utility of applying the CSP formalization to conflict resolution by pointing out where the proposed technique is expected to be most effective. At the end of the chapter we compare our method with other related work in conflict resolution.

7.2 Conflict Resolution Constraints

7.2.1 Conflicts and Conflict Resolution Methods

In Chapter 2 we described two kinds of threats, positive and negative. We referred to the negative threats as "conflicts," since they pose immediate danger to the correctness of a plan. In this section we we develop a representation for conflict resolution.

Let $cl = \langle \mathrm{s}_i \xrightarrow{p} \mathrm{s}_j \rangle$ be a causal link in a plan, and $t = \langle \mathrm{s}_k, e \rangle$, where e is an effect of s_k, be a negative threat to cl (See Chapter 2 for a definition of negative threat). A conflict is defined as a pair `conf(cl, t)`. We will refer to s_i as a *producer* of p, s_j as the *user* of p, s_k as a clobberer and q as the clobbering effect.

Suppose that the ordering relation in a plan Π is represented using a transitive closure; let $TR(R)$ denote the transitive closure of a precedence relation R. Then for a given causal link cl in a plan, the set of all conflicts for cl can be found in time $O(n)$, where n is the total number of steps in the plan.

For each conflict $\texttt{Conf} = \texttt{conf}(\langle \mathrm{s}_i \xrightarrow{p} \mathrm{s}_j \rangle, \langle \mathrm{s}_k, e \rangle)$, let $M(\texttt{Conf})$ be the set of all resolution methods for resolving `Conf`. The set of alternative methods can be represented as a disjunctive set:

Table 7.1. Conflicts in the painting example. All conflicts have the same producer operator $s_i = s_{init}$

Conflict	Precondition	User	Clobberer	Clobbering Effect
$\mathtt{Conf}_a$	Handempty	get-brush(?*cb*)	get-brush(?*lb*)	¬Handempty
$\mathtt{Conf}_b$	Handempty	get-brush(?*lb*)	get-brush(?*cb*)	¬Handempty
$\mathtt{Conf}_c$	Dry(?*cb*)	get-brush(?*cb*)	paint ladder	¬Dry(?*lb*)
$\mathtt{Conf}_d$	Dry(?*lb*)	get-brush(?*lb*)	paint-ceiling	¬Dry(?*cb*)
$\mathtt{Conf}_e$	Dry(Ladder)	paint-ceiling	paint-ladder	¬ Dry(Ladder)

$$M(\mathtt{Conf}) = \{\{s_j \mapsto s_k\}, \{s_k \mapsto s_i\}, \{e \neq \neg p\}\}$$

In actual implementation for generating the constraints, however, the total number of conflict-resolution methods can be reduced by taking into account the structure of the plan. For example, if the three operators s_i, s_j, and s_k are ordered in a linear sequence in a plan, such that the threat s_k is located necessarily between s_i and s_j, then only the separation methods are applicable for resolving the conflict without violating the existing ordering constraints.

Consider the painting example introduced earlier. A plan for painting both the ceiling and the ladder consists of two unordered linear sequences of operators:

$s_{init}\mapsto$**get-brush**(?*cb*)$\mapsto$**paint-ceiling**(Ceiling)
$\quad\mapsto$**return-brush**(?*cb*)$\mapsto s_{finish}$
$s_{init}\mapsto$**get-brush**(?*lb*)$\mapsto$**paint-ladder**(Ladder)
$\quad\mapsto$**return-brush**(?*lb*)$\mapsto s_{finish}$

In the above, ?*cb* is a variable which stands for any ceiling brush, and ?*lb* is a variable that likewise refers to any ladder brush. The preconditions and effects of each operator in the plan are shown in Table 2.1, in Chapter 2. The conflicts in this plan are listed in Table 7.1.

The conflict resolution methods are:

M($\mathtt{Conf}_a$)= {{**get-brush**(?*cb*)$\mapsto$**get-brush**(?*lb*)},
{**return-brush**(?*lb*)$\mapsto$**get-brush**(?*cb*)}}.
M($\mathtt{Conf}_b$)= {{**get-brush**(?*lb*)$\mapsto$**get-brush**(?*cb*)},
{**return-brush**(?*cb*)$\mapsto$**get-brush**(?*lb*)}}.
M($\mathtt{Conf}_c$)={{**get-brush**(?*cb*)$\mapsto$**paint-ladder**}, {?*cb* ≠?*lb*}}.
M($\mathtt{Conf}_d$)={{**get-brush**(?*lb*)$\mapsto$**paint-ceiling**}, {?*cb* ≠?*lb*}}.
M($\mathtt{Conf}_e$)={{**paint-ceiling**$\mapsto$**paint-ladder**}}.

7.2.2 Relations Among Conflict Resolution Methods

To find one or all of the consistent methods for resolving a set of conflicts in a plan Π, one has to take into account the various kinds of relationships among different constraints. In this section, we define and analyze two types of relation, the inconsistency relation and the subsumption relation.

Inconsistency Relation

Resolving conflicts involves imposing constraints onto the structure of a plan. Some constraints cannot be imposed together, because they are *inconsistent* with each other. For example, imposing two ordering constraints $\alpha_1 \mapsto \alpha_2$ and $\alpha_2 \mapsto \alpha_1$ onto the same plan results in a cycle in step ordering, which is disallowed in a partial order. Likewise, constraints $?x_1 = ?x_2$ and $?x_2 \neq ?x_3$ are inconsistent if the variables are already constrained in the plan such that $?x_1 = ?x_3$.

With the transitive-closure representation TR for the step-ordering relation, inconsistency can be easily defined. Let R_1 and R_2 be two sets of ordering constraints. R_1 *is inconsistent with* R_2 in a plan Π, or $I_\Pi(R_1, R_2)$, if and only if for some plan step $\mathbf{s}$,

$$(\mathbf{s}, \mathbf{s}) \in TR(\text{Order}(\Pi) \cup R_1 \cup R_2).$$

That is, R_1 and R_2 are inconsistent if and only if imposing both onto the plan steps would result in a cycle.

We can similarly define an inconsistency relation among two variable-binding constraints. Let R_1 and R_2 be two sets of variable-binding constraints. These constraints are either of the form $?x = ?y$, or of the form $?x \neq ?z$. The variable-binding constraints $?x = ?y$ in both R_1 and R_2 form an *equivalence relation* on the plan variables. An equivalence relation R on a set S is defined as follows:

- R *is reflexive;* that is, for every member a of R, $(a, a) \in R$.
- R *is symmetric;* that is, for every pair of members a and b of R, if $(a, b) \in R$ then $(b, a) \in R$.
- R *is transitive;* that is, for all members a, b and c in R, if $(a, b) \in R$ and $(b, c) \in R$ then $(a, c) \in R$.

The first requirement in the above definition states that for every variable $?x$, $?x = ?x$ always holds. The transitivity requirement states that for all plan variables $?x_1$, $?x_2$ and $?x_3$, if $?x_1 = ?x_2$ and $?x_2 = ?x_3$, then $?x_1 = ?x_3$. Below, we consider the constraint $=$ to form an equivalence relation.

Consider two variable-binding constraint sets R_1 and R_2. We say that, R_1 is *inconsistent* with R_2 in a plan Π, or $I_\Pi(R_1, R_2)$, if and only if there exists a plan variable $?x$, such that

$$?x \neq ?x \in \text{Binding}(\Pi) \cup R_1 \cup R_2.$$

Two constraints, ordering or variable-binding, are *consistent* if they are not inconsistent with each other.

Subsumption Relation

Imposing one set of constraints may make another set redundant. For example, let R_1 be $\{s_2 \mapsto s_3\}$, and R_2 be $\{s_1 \mapsto s_3\}$. If $(s_1 \mapsto s_2) \in \text{Order}(\Pi)$, then imposing R_1 would make it unnecessary to further impose R_2.

In general, R_1 subsumes R_2 if imposing R_1 will guarantee that R_2 is also imposed. Thus, R_2 is considered to be weaker than R_1. More precisely, let R_1 and R_2 be two sets of conjunctive ordering constraints. R_1 *subsumes* R_2 *in plan* Π, or $S_\Pi(R_1, R_2)$, if and only if

$$R_2 \subseteq TR(R_1 \cup \text{Order}(\Pi)).$$

Let R_1 and R_2 be two sets of variable-binding constraints. R_1 *subsumes* R_2 *in plan* Π, or $S_\Pi(R_1, R_2)$, if and only if

1. every variable-binding constraint $?x = ?y$ of R_2 is a member of R_1, and
2. every separation constraint $?u \neq ?v$ of R_2 can be inferred from R_1 and Π. That is, for some variables $?x$ and $?y$ such that $?x = ?u$, and $?y = ?v$, we have

$$(?x \neq ?y) \in R_1 \cup \text{Binding}(\Pi)$$

As an example, let $R_1 = \{(?x = ?y)\}$ and $R_2 = \{(?y \neq ?z)\}$. If $(?x \neq ?z) \in \text{Binding}(\Pi)$ then $S_\Pi(R_1, R_2)$ holds.

Given a plan Π, one can establish the subsumption and inconsistency relations between any pair of constraints R_1 and R_2, by computing the transitive and equivalence closures of the ordering and variable-binding constraints, respectively. The former takes $O(n^3)$ time to compute for a plan with n steps, while the latter takes $O(v^3)$ time for a plan with v variables.

For convenience, the subscript Π used for both I_Π and S_Π relations is dropped in situations where it is clear about the plan under consideration.

Because the subsumption relation S is defined via subset relations, it can be easily verified that S is transitive. That is,

Lemma 7.2.1 *If* $S(R_1, R_2)$ *and* $S(R_2, R_3)$, *then* $S(R_1, R_3)$.

In addition, it is also easy to see that the following property holds:

Lemma 7.2.2 *If $S(R_1, R_2)$, $S(R_3, R_4)$ and $I(R_2, R_4)$, then $I(R_1, R_3)$.*

This lemma states that if two constraints are inconsistent, then any stronger versions of the two constraints are also inconsistent. By letting R_3 and R_4 both be R', it holds as a corollary that if $S(R_1, R_2)$, and $I(R_2, R')$, then $I(R_1, R')$.

Subsumption and inconsistency relations have so far been defined between a pair of constraints. Extensions to higher-order relations can be made in a similar manner. For example, two sets of precedence constraints R_1 and R_2 subsume R_3 in a plan Π if and only if $R_3 \subseteq TR(R_1 \cup R_2 \cup \text{Order}(\Pi))$.

Minimal Solutions

If $\mathcal{C}$ is the set of all conflicts in plan Π, and if a consistent set of constraints *Sol* resolves all conflicts in Π, then *Sol* is called a *solution* to $\mathcal{C}$.

It is possible to find a certain amount of redundancy in a solution. For example, suppose that in a plan Π there is only one conflict, which could be resolved by either $\mathbf{s}_1 \mapsto \mathbf{s}_2$ or $?x \neq ?y$. The stronger constraint $Sol = \{(\mathbf{s}_1 \mapsto \mathbf{s}_2), (?x \neq ?y)\}$ also resolves `Conf`. However, the latter is unnecessarily strong, because either conjunct is able to resolve the conflict without the other. In this case, it is possible to reduce the solution *Sol* to a weaker one which is in some sense *minimal*. A *minimal solution* Sol_{min} for conflicts $\mathcal{C}$ is a solution for $\mathcal{C}$ such that no proper subset of Sol_{min} resolves all conflicts in $\mathcal{C}$.

As the previous example illustrates, a solution may have several alternative sets of minimal solutions; in the simple example they are $sol_1 = \{\mathbf{s}_1 \mapsto \mathbf{s}_2\}$ and $sol_2 = \{?x \neq ?y\}$. If Sol' is a minimal solution, and if $Sol' \subset Sol$, then the constraints in the set difference $(Sol - Sol')$ are considered redundant with respect to Sol'. As a consequence, if R_1 and R_2 are two disjoint subsets of a solution *Sol*, and if R_1 subsumes R_2 (i.e., $S(R_1, R_2)$), then removing R_2 from *Sol* leaves at least one minimal solution *intact*. This observation is the basis of a constraint-propagation rule presented in the next section.

7.3 Conflict Resolution as Constraint Satisfaction

In Chapter 4 we introduced constraint satisfaction as a mathematical and algorithmic tool for solving a class of combinatorial problems. In this section we will model conflict resolution as constraint satisfaction.

7.3.1 Representation

Conflict resolution in planning can be modeled by a CSP. For a given plan Π, a conflict $\texttt{Conf}_i$ in Π corresponds to a variable. The domain of $\texttt{Conf}_i$ is the set of alternative conflict-resolution methods that are capable of resolving the conflict. The constraints among the variables are defined via the consistency relations among different sets of conflict resolution methods. A solution to the CSP corresponds to a set of consistent resolution methods that resolves all conflicts in Π.

An advantage of the mapping from conflict resolution problems to CSPs is that many existing strategies for solving general CSPs can be directly applied to facilitate a global analysis of conflicts. In addition, the existence of subsumption relations among the variables provides new opportunities for simplifying a CSP further than permitted by traditional CSP techniques.

The methods for solving a CSP can be roughly divided into two categories: local constraint propagation and global, heuristically-guided backtracking algorithms.

7.3.2 Propagating Constraints Among Conflicts

When two or more variables are considered together, certain implicit constraints among them can be inferred from the given ones. Consider, for example, a plan containing two conflicts, $\texttt{Conf}_A$ and $\texttt{Conf}_B$, where the resolution methods have been found out to be

$$M(\texttt{Conf}_A) = \{\{x = y, b \mapsto c\}, \ldots\}, \text{ and } M(\texttt{Conf}_B) = \{\{x \neq y\}, \{c \mapsto b\}\}.$$

From the inconsistency relation $I(x = y, x \neq y)$ and $I(b \mapsto c, c \mapsto b)$, it is clear that the constraint set $\{x = y, b \mapsto c\}$, for $\texttt{Conf}_A$, cannot be used as part of a solution for solving both conflicts. Therefore, it can be removed from the set of resolution methods for $\texttt{Conf}_a$ without affecting any solution. Furthermore, if it is also the only method for resolving $\texttt{Conf}_a$, then the plan Π corresponds to a dead end; it cannot be resolved by simply imposing constraints.

The above example is an instance of a general procedure known as *arc-consistency* for solving a CSP; Chapter 4 gives a systematic review for these and other CSP techniques.

7.3.3 Redundancy Removal via Subsumption Relations

Recall that a conflict-resolution method R_1 subsumes R_2, if R_1 is stronger than R_2. The presence of this relationship makes certain constraints redundant. In this section, we consider two cases in which redundancy can be detected.

Redundant Conflicts. Consider a plan Π containing, among others, two conflicts, $\texttt{Conf}_1$ and $\texttt{Conf}_2$. Suppose that every set of constraints for $\texttt{Conf}_1$ subsumes *some* constraint for $\texttt{Conf}_2$. Because any solution for resolving all conflicts in the plan must resolve $\texttt{Conf}_1$, every choice of a resolution method from $M(\texttt{Conf}_1)$ must also resolve $\texttt{Conf}_2$. This fact holds even when a constraint chosen from $M(\texttt{Conf}_2)$ is removed from *Sol*. Therefore, if the conflict

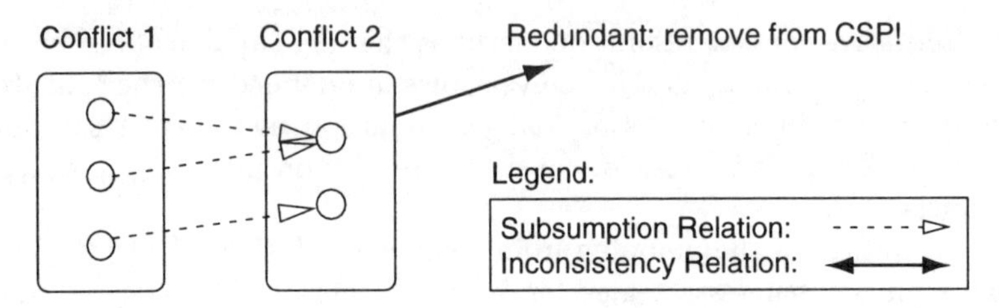

Fig. 7.1. Conflict 2 can be removed from a CSP by Var-Subsumption Theorem

$\texttt{Conf}_2$ is removed from further consideration when solving the CSP, any minimal solution to the modified CSP could still solve the original CSP. The argument is summarized in the theorem below.

Theorem 7.3.1 (Var-Subsumption). *Let Π be a plan with a conflict set $\mathcal{C}$. Let $\texttt{Conf}_1$ and $\texttt{Conf}_2$ be two conflicts in $\mathcal{C}$. Let $M(\texttt{Conf}_1)$ and $M(\texttt{Conf}_2)$ be conflict resolution constraints for $\texttt{Conf}_1$ and $\texttt{Conf}_2$, respectively.*

Suppose that

$$\forall R_1 \in M(\texttt{Conf}_1), \exists R_2 \in M(\texttt{Conf}_2) \textit{ such that } S(R_1, R_2)$$

Then $\texttt{Conf}_2$ can be pruned from the CSP without affecting any minimal solution for $\mathcal{C}$.

In this case, we say that $\texttt{Conf}_2$ is redundant. The theorem is illustrated in Figure 7.1.

Pruning of redundant variables from a CSP reduces the size of the CSP and therefore can lead to improved efficiency in constraint reasoning.

Redundant Constraints. Removal of redundancy in the above form only utilizes the subsumption information. When both inconsistency and subsumption relations are considered together, it is also possible to remove individual redundant values from the domain of a CSP variable.

Consider again a plan Π containing two conflicts, $\texttt{Conf}_1$ and $\texttt{Conf}_2$. Suppose that there is some constraint set R_2 in $M(\texttt{Conf}_2)$, such that for every method R_1 in $M(\texttt{Conf}_1)$, either

1. $I(R_1, R_2)$, i.e., R_1 is inconsistent with R_2; or
2. $\exists R_3 \in \texttt{Conf}_2$ where $R_3 \neq R_2$, such that $S(R_1, R_3)$. That is, R_1 subsumes some other constraints in $M(\texttt{Conf}_2)$.

Given the above conditions, we could show that R_2 is redundant. To see this, consider a solution for the CSP. In this solution there must be a constraint for solving $\texttt{Conf}_1$. Let this constraint be R_1. If R_1 is inconsistent with R_2, then R_2 cannot be part of the solution. If, on the other hand, one can find an R_3 in $M(\texttt{Conf}_2)$ such that R_1 subsumes R_3, then R_2 need not be part of $M(\texttt{Conf}_2)$ since R_1 alone can resolve $\texttt{Conf}_2$. As a result, no matter what method is chosen for $\texttt{Conf}_1$, R_2 will not be chosen for a minimal solution.

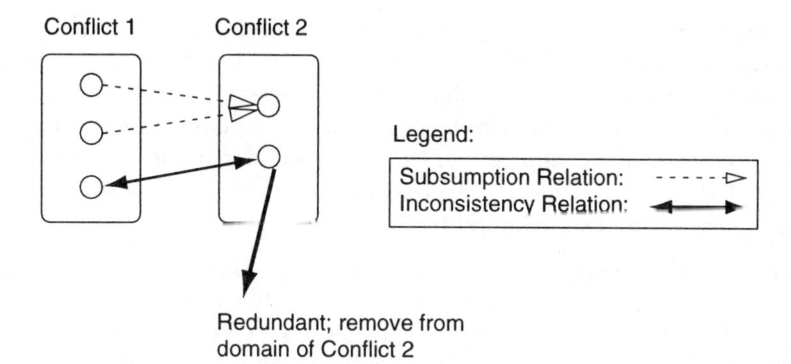

Fig. 7.2. A conflict-resolution constraint of Conflict 2 can be removed from the domain of Conflict 2 by Value-Subsumption Theorem

This means that R_2 can be removed from $M(\texttt{Conf}_2)$ without affecting the set of minimal solutions for resolving all conflicts in Π. This conclusion is summarized in the following theorem.

Theorem 7.3.2 (Value-Subsumption). *Let Π be a plan with a conflict set $\mathcal{C}$. Let $\texttt{Conf}_1$ and $\texttt{Conf}_2$ be two conflicts in $\mathcal{C}$, and let $M(\texttt{Conf}_1)$ and $M(\texttt{Conf}_2)$ be their corresponding sets of conflict-resolution constraints. Suppose that there exists some R_2 in $M(\texttt{Conf}_2)$, such that for all R_1 in $M(\texttt{Conf}_1)$, either*

1. *$I(R_1, R_2)$, or*
2. *$\exists R_3 \in M(\texttt{Conf}_2)$. $R_3 \neq R_2$ and $S(R_1, R_3)$.*

Then R_2 can be pruned from $M(\texttt{Conf}_2)$ without affecting the set of minimal solutions to $\mathcal{C}$.

This theorem is shown pictorially in Figure 7.2.

Augmenting Arc-Consistency. Removal of redundant variables or values in a CSP is called *redundancy removal.* It can be used to augment a traditional arc-consistency algorithm, described in Chapter 4, in the following manner: every time a pair of variables $(X,\ Y)$ is examined in an arc-consistency algorithm, a check is made to verify whether the variable Y is redundant using the Var-Subsumption Theorem (Theorem 7.3.1). The variable can be removed if the test returns true.

A second test, using the Value-Subsumption Theorem (Theorem 7.3.2), can be performed to check whether a value of Y is redundant. Every redundant value v can be removed from the domain of variable Y.

For a given set of constraints, the computations of both inconsistency and subsumption relations have the same time complexity. These relations are computed only once when a CSP is first initialized. Furthermore, the augmented arc-consistency algorithm takes these two relations as inputs, and considers pairs of conflicts for both of them. Therefore, the additional consideration of subsumption relations in the augmented algorithm increases the complexity of the original algorithm only by a constant factor.

Redundancy removal can also be extended in a similar manner to augment path-consistency algorithms, which examine and infer inconsistencies within groups of three variables [91]. Such an extension is straightforward, and a detailed description can be found in [153].

Augmenting Forward-Checking. Forward-checking is a form of combining partial arc-consistency and backtracking. A full description is given in Chapter 4. Here we discuss how to augment a key subroutine, FWDREVISE, by taking into account the additional subsumption information.

Recall that FWDREVISE takes two arguments, a current variable assignment $X = u$, and a set representing the remaining CSP. FWDREVISE performs constraint propagation through the un-assigned variables, by considering pairs of the form $(X = u, \langle Y, \{v1, v2, \ldots vn\}\rangle)$. In this pair Y is a variable in CSP, and the set contains Y's domain values. FWDREVISE checks each value of Y in turn:

for each $\langle Y, M(Y)\rangle \in$ CSP **do**
  if there is a value $v \in M(Y)$ such that (u, v) is inconsistent,
    then delete v from $M(Y)$;
  end if;
  if there is a value $v \in M(Y)$ such that u subsumes v,
    then delete Y from CSP;
  end if;
end for.

7.4 The Painting Example

Consider again the painting problem described in Section 7.1. The conflict-resolution methods for resolving all conflicts in this problem have been formulated in Section 7.2. To facilitate understanding, we rename the conflict-resolution methods as follows:

M(Conf$_a$)= $\{a_1,\ a_2\}$, M(Conf$_b$)= $\{b_1,\ b_2\}$,
M(Conf$_c$)= $\{c_1,\ c_2\}$, M(Conf$_d$)= $\{d_1,\ d_2\}$,
M(Conf$_e$)=$\{e_1\}$.

In the above, a_1 corresponds to the first conflict resolution method for the conflict Conf$_a$, and the rest are likewise defined. These conflicts are converted to a constraint satisfaction problem, shown in Figure 7.3.

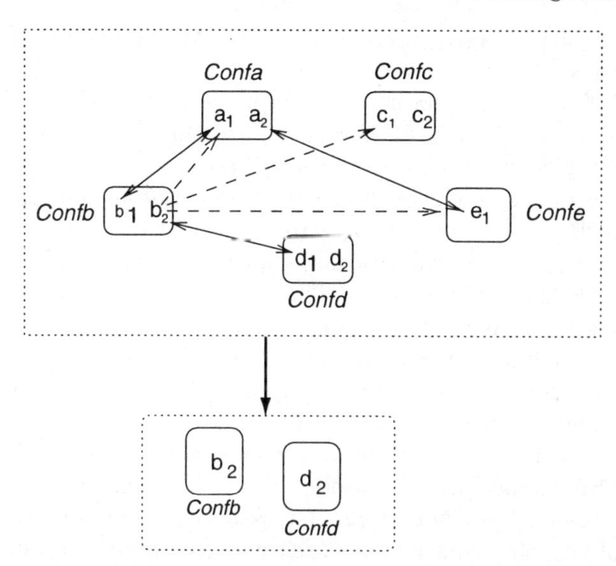

Fig. 7.3. A CSP for the conflict-resolution in the painting domain. The solid arrows correspond to inconsistency relations, and the dashed ones the subsumption relations

The conflict resolution process is as the following:

(1) We start with $\texttt{Conf}_e$ which has the smallest cardinality. The only choice for $\texttt{Conf}_e$ is inconsistent with the second constraint set, a_2, for $\texttt{Conf}_a$. Therefore, the latter is removed from M($\texttt{Conf}_a$) by arc-consistency.
(2) Next, the only alternative left for M($\texttt{Conf}_a$) is $\{a_1\}$, which is inconsistent with the first choice for $\texttt{Conf}_b$. Thus, due to arc-consistency M($\texttt{Conf}_b$) is reduced to $\{b_2\}$.
(3) The currently remaining constraint for $\texttt{Conf}_b$ subsumes some constraints for resolving $\texttt{Conf}_a$, $\texttt{Conf}_c$, and $\texttt{Conf}_e$. Thus, from the Var-Subsumption Theorem, all three conflicts become redundant and thus can be removed.
(4) The remaining value $\{b_2\}$ for $\texttt{Conf}_b$ is inconsistent with the first constraint for $\texttt{Conf}_d$. Therefore the latter can be removed by arc-consistency.
(5) Finally, the CSP contains only two conflicts, $\texttt{Conf}_b$ and $\texttt{Conf}_d$, and the remaining constraints left in $M(\texttt{Conf}_b)$ and $M(\texttt{Conf}_d)$ are consistent with each other, and can be combined as a global solution to the CSP:

$$\texttt{return-brush}(?cb) \mapsto \texttt{get-brush}(?lb), ?cb \neq ?lb$$

This solution is also a minimal solution. The resulting plan is formed by ordering all ceiling-painting operations to be before all ladder-painting operations, and by making sure that the ceiling-painting brush is different from the ladder-painting brush.

7.5 When Is the Constraint-Based Method Useful?

Basic planning methods introduced in Chapter 2 have dealt with the problem of conflict resolution in a local, incremental manner. In each planning iteration only a few conflicts are introduced. In these systems, constraint-based conflict resolution may not show its greatest potential. In contrast, we expect the global conflict-resolution technique to be most useful when a large number of conflicts exist, and when an arbitrary choice in conflict resolution methods is not likely to lead to a final solution. In a problem-decomposition framework, many conflicts are likely to occur at once when solution plans to different subproblems are combined. Therefore, our constraint-based conflict-resolution technique is expected to be particularly useful for combining plans. Below, we consider a few other uses of this theory.

Related to problem-decomposition systems, a second type of planner for which the CSP method may be useful is one that employs a task-network hierarchy introduced in Chapter 12. A task-network planning system starts with a set of subgoals, and reduces each one according to a library of predefined networks of sub-plans. Each sub-plan may also contain more detailed subgoals that can be further reduced. The system terminates when every remaining operator in the plan can be successfully executed. Examples of such systems are SIPE [147], NONLIN [132], and DEVISER [140]. When these systems are applied to complex domains, each task reduction may introduce many new steps that interact, which may in turn cause a large number of conflicts to occur.

In addition, the CSP method is expected to be useful in *plan revision*, where the input is a used plan that is possibly incorrect, and the output is a modified version of the plan which fits a new situation. For example, in the PRIAR system of Kambhampati and Hendler [70], a previously generated plan is first retrieved. The system then identifies those preconditions of the operators that are no longer established or conflicted in a new situation, and proposes plan-modification operations for re-achieving them. The inserted new operators may render the plan possibly incorrect by creating new conflicts, and the number of such conflicts may increase with the number of faults in the original plan. In this case, our CSP method for conflict resolution can fix the remaining conflicts and arrive quickly at a conclusion about the validity of the fix.

7.6 The COMBINE Algorithm

This section describes an implementation of the above conflict resolution techniques in the COMBINE algorithm. The input of COMBINE is assumed to be a possibly incorrect plan, that is, a partial-order plan for which some instance might be incorrect. The algorithm outputs a correct partial-order plan derived from the input plan, if a correct plan exists.

Table 7.2. The COMBINE algorithm

Algorithm COMBINE(Π)
Input: A plan Π.
External Functions:

- FWDCHECKING(CSP) is the forward-checking algorithm, described in Chapter 4.
- PARTIALARC(CSP) returns a CSP as a result of a partial arc-consistency operation.
- BUILDLINKS(Π) returns the plan with all causal links established.
- DETECTCONFS(cl,Π) returns the set of all conflicts with a causal link cl in a plan Π.
- BUILDCSP(cl,Π) constructs a CSP representation of conflict-resolution constraints.
- APPLYCONSTRAINTS(Solution,Π) returns Π upon which all constraints in Solution are imposed.

Output: A correct version of Π, if one exists; Fail otherwise.

```
1.   if Π does not already have causal links then
2.      Π := BuildLinks(Π);
3.   end if
4.   Conflicts := ∅;
5.   for each causal link cl in C-Links(Π) do
6.      Conflicts := Conflicts∪DetectConfs(cl, Π);
7.   end for
8.   CSP := BuildCSP(Conflicts, Π);
9.   CSP := PartialArc(CSP);
10.  if the domain of a variable in CSP becomes empty, then
11.     return(Fail);
12.  end if
13.  Solution := FwdChecking(CSP);
14.  if Solution ≠ Fail then
15.     Π := ApplyConstraints(Solution, Π);
16.     return(Π);
17.  else return(Fail);
18.  end if
```

The COMBINE algorithm is shown in Table 7.2. The first task of COMBINE is to detect all conflicts with the causal links in the input plan. If the input plan does not come with causal links, then COMBINE will find the causal links in the plan by examining the producer-user relations (subroutine BUILDLINKS, Step 2). It does this by iterating through the preconditions in the plan, and for each precondition, it will look for a producer to establish it. If there are several alternative choices in producers, then one of them is chosen and the rest are stored as backtrack points.

After all conflicts are detected and recorded, COMBINE proposes a set of conflict-resolution methods as outlined in Section 2. The conflicts and the conflict-resolution methods form the basis for a CSP-representation. This is done in Step 8, in procedure BUILDCSP.

Once the CSP-representation is obtained, a partial arc-consistency algorithm is applied. This procedure performs two tasks:

- using a partial arc-consistency algorithm to check for dead ends and to remove inconsistencies, and
- using a redundancy-removal algorithm for eliminating redundant conflicts or constraints in the CSP.

The partial arc-consistency algorithm checks, for every pair X and Y of variables in the CSP, whether a value of X is inconsistent with every value of Y. If so, then the value is removed from the domain of X. This is essentially the REVISE algorithm of AC, which is described in Table 4.6. A difference from AC is that here we only perform REVISE once through the constraint-network. Thus, it doesn't re-check the consistency of X with other variables after an update in X's domain. Although the partial arc-consistency algorithm does not ensure that the CSP network be completely arc-consistent, it does allow a significant number of inconsistencies to be removed. This implementation decision is based on the need to minimize the complexity of preprocessing algorithms.

After partial arc-consistency is performed, a redundancy-removal procedure is followed. This procedure checks every pair of variables X and Y, to see if every value of X subsumes some value of Y. By the Var-Subsumption Theorem, Y can be removed from the CSP if the condition is true. If so, then Y can be removed from the CSP. The Value-Subsumption Theorem is applied similarly for removal of redundant values. In this process, a count is maintained for the total number of times in which a value u of a variable X subsumes some value v of other variables in the CSP. The recorded measure $s(u)$ will be used in the forward-checking as part of a variable-ordering heuristic.

Finally, algorithm FWDCHECKING will be executed to find a set of consistent constraints for removing all conflicts. A subroutine of FWDCHECKING is VARORDERING(CSP), which sorts the variables of the CSP in the ascending order of their domain sizes. For variables of the same domain size, an option can be selected to order them in decreasing subsumption measure $s(u)$, computed in the previous step. The purpose of this ordering process is for the backtracking algorithm to search a smaller search tree. During a backtracking process, an un-assigned variable can be removed from the remaining CSP if one of its values is subsumed by a previously assigned value. This enables the removal of redundant nodes as quickly as possible.

When a consistent solution is found, it will be applied to Π to generate a correct solution plan (Step 15).

To evaluate COMBINE, we designed an artificially generated plan set with conflicts, on which we ran both COMBINE and a basic partial-order planner

TWEAK (see Chapter 3) to compare their performance. This experiment is indicative of both systems when they are applied to the problem of plan-revision. The experimental results are discussed in detail in [154]. Here we summarize the main results:

1. The COMBINE algorithm easily outperforms TWEAK in time in this domain. The performance difference increases as the number of conflicts in a plan grows.
2. For plans containing conflicts that are not resolvable, COMBINE can determine that all resolution methods lead to failure more quickly than TWEAK. On the other hand, TWEAK has to go though an entire planning iteration and could not utilize an intelligent order in which to resolve conflicts.
3. The subsumption relation is more useful when conflicts are tightly coupled in a small portion of a plan, as indicated by the painting example.

7.7 Related Work

7.7.1 Overview of Previous Work

Because of its importance in planning, methods for conflict resolution were explored early on. Sussman's system HACKER [130] recognized and fixed "bugs," which were certain classes of conflicts. The bugs were fixed using a bag of hacks, which were different types of ordering constraints that could be imposed upon the operators of a plan. Sacerdoti's NOAH [114] used a partial order to represent the structure of a plan, and implemented a more elaborate set of ordering constraints that could resolve different classes of conflicts. A problem with NOAH is that given several alternative choices in the constraints, it commits to one of them, and does not have the ability to backtrack should an inconsistent situation occur later. Tate's NONLIN [132] fixed this problem and introduced a complete set of alternative ordering constraints that are capable of resolving conflicts in a completely instantiated plan. Recognizing the need to represent resources using variables in a plan, Stefik's MOLGEN [128] and Wilkins' SIPE [147] both could further impose constraints on variable bindings to resolve conflicts. Chapman's TWEAK [28] introduced a white-knight constraint for resolving conflicts. Extending the operator language to more elaborate forms, Barrett and Weld [17] used *confrontation* to address actions with conditional effects.

A major theme of the previous approaches to conflict reasoning is that conflicts are resolved incrementally and locally. In these systems, conflicts are reasoned about one at a time. In the presence of more than one conflict, no systematic theory exists to guide the conflict-resolution process. A drawback then is their loss of computational efficiency. A notable exception is the work by Joslin and Pollack [69], who take constraint-based conflict resolution as the basis of their algorithm.

7.7.2 Related Work by Smith and Peot

A related work on conflict-resolution is that by Smith and Peot, who proposed a technique to postpone threat/conflict-resolution in a basic backtrack-chaining, partial-order planner. Their key observation is that certain threats, or groups of threats, in a plan do not have to be resolved as soon as they appear. The consideration of these threats can be delayed without sacrificing the possibilities of finding a solution. They argue that doing so has the advantage of leaving a plan less constrained, enabling a planner to take more advantage of its partial-ordered nature.

In this section, we show that the threat-analysis technique of Smith and Peot is in fact consistent with the formalization of conflict-resolution as constraint satisfaction. Some of their threat-removal results are in fact derivable from arc-consistency properties, others are consistency properties relating N CSP-variables. To begin with, we first review their key results [124].

A fundamental concept in [124] is an *operator graph.* This is a directed graph representing the inter-relation between planning operators. In the graph there are two kinds of nodes, a *precondition node* for each precondition, and an *operator node* for every operator. Between a precondition node and an operator node there is a directed edge if the precondition is required by the operator. Similarly, if an operator has an effect which could unify with a precondition, then there is a directed edge from the former to the latter. In addition, each operator occurs only once in the graph.

To illustrate, Figure 7.4 shows an operator graph for a simplified painting domain. In an operator graph, an operator node α threatens a precondition node p if some effect of α unifies with the negation of p. By a convention used by Smith and Peot, a boldfaced arc denotes a threat (see Figure 7.4).

Threats in an operator graph can be removed by imposing ordering constraints between operators (for simplicity, in their exposition they temporarily ignored variable-binding constraints). In the example of Figure 7.5, adopted from [124], the threat between nodes 2 and B can be delayed because, no matter how the rest of the threats in a plan are resolved, there always remains a choice in which the operator node 2 can be ordered before operator node 1, which resolves the threat.

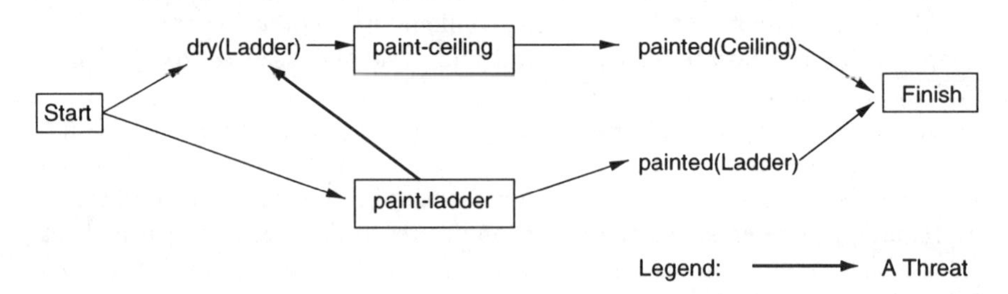

Fig. 7.4. An operator graph for a simplified painting domain

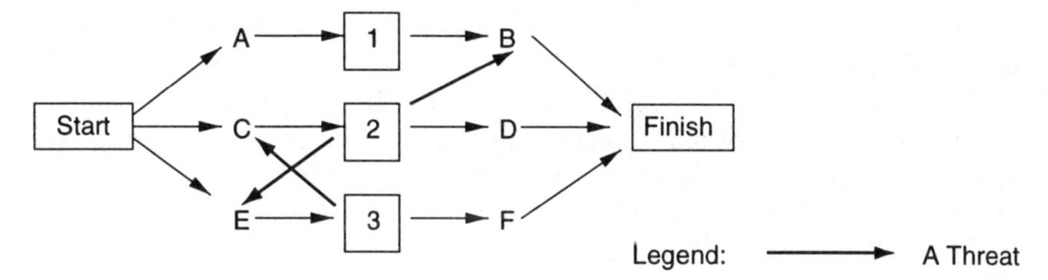

Fig. 7.5. An operator graph showing how a threat, from node 2 to node B, can be delayed

This observation is generalized by Smith and Peot as follows: Let T be the set of threats in an operator graph, and let P be a subset of those threats. The threats in P can be postponed if there is a set of ordering constraints that resolves the threats in P for every possible resolution of the remaining threats in T-P.

This result is in fact a direct generalization of arc-consistency in a CSP. Consider the definition for an arc-consistency relation between two variables in a CSP. Recall that a pair of variables (X, Y) is arc-consistent if there is a consistent value in the domain of Y for every possible value of X. It is not hard to observe the parallel between the above theorem and the arc-consistency definition. In our formalization of conflict resolution as a CSP, every threat is a variable, and the threat-removal constraints are the variables' values. If we use the structure of an operator-graph to define a consistency relation on operator ordering, such that no operator α is ordered both before and after another operator β, then we have a CSP representation. A solution to the CSP is a set of consistent ordering constraints that could resolve all conflicts in the operator graph. From this view, we could obtain the same conclusion as that by Smith and Peot.

In Constraint Satisfaction area, Freuder [53] has shown that when (a) the constraint relation in a CSP is binary, (b) the constraint network is a tree, and (c) the variable pairs are arc-consistent, the CSP can be solved in a *backtrack-free* order (see Theorem 4.5.1 in Chapter 4). This backtrack-free order is one in which the variables are instantiated, such that X is instantiated before Y only if (X,Y) is arc-consistent.

Applying this result to the threat-removal problem studied by Smith and Peot, we observe the following. Let (T-P) represent an aggregate variable in a CSP, and P represent another. An aggregate variable is formed by representing all threats in the set (T-P) as a single variable, and the Cartesian product of all constraints, as long as they are consistent together, as the values of the variable. Then Freuder's theorem can be directly applied: The CSP is obviously binary since there are only two variables. For the same reason, it is a tree network of variables. Finally, we assume that this CSP is arc-consistent; every value of (T-P) leaves a consistent value for P. Then by Freuder's backtrack-free order, (T-P) should be solved before P — this is essentially the result of Smith and Peot as we described above.

From the above discussion, we see that the CSP formalization of conflict-resolution problems not only offers computational advantages, but also serves as a unifying framework in which some related work in this area, such as the one by Smith and Peot, could be understood from a new angle.

7.7.3 Related Work by Etzioni

Etzioni studied conflicts in the context of a total-order planner PRODIGY [25]. In his STATIC system, the operator definitions are compiled into an AND/OR tree. Based on this AND/OR tree STATIC extracts useful information about necessary subgoal clobberings and compiles this information into search-control rules. The rules are then used during the operation of a total-order planner to sequence the goals in order to avoid certain predictable subgoal interactions. It was shown that these rules are as effective as those learned during planning by explanation-based learning (EBL) systems such as PRODIGY.

A difference between STATIC and our framework here is that STATIC computes all *necessary* interactions between pairs of subgoals. These are the interactions that are guaranteed to occur as long as the two subgoals appear simultaneously in a plan. In our case, however, we compute all *possible* conflicts among subgoals, anticipating what might occur during planning. These conflicts are used globally to obtain a conflict resolution strategy.

7.8 Open Problems

Temporal Operator Language

One advantage of our theory is its extensibility; with a more elaborate planning language, the underlying theory for global conflict resolution need not change. For example, one can extend the partial-order planning language to include the time-point algebra of Vilain and Kautz [141], by associating the occurrence of each action with a time point. One can also extend the partial-order planning language to include Allen's interval representation of actions. With Vilain and Kautz's time-point logic, the relationships between two time points include "precedes," "follows," "same," and "not-same." With the new language, one can also augment the set of conflict resolution methods by providing an additional set of constraints. For example, suppose that whenever two operators occur simultaneously, one of their combined effects will clobber an establishment relation. Then one way to resolve the conflict is to impose a "not-same" constraint onto the time points of the two operators. This augmentation only enlarges the domain of individual variables that represent conflicts in a CSP, and thus the same computational framework can be directly applied to resolve conflicts in the extended language.

Domain-Dependent Conflict Resolution

So far we have considered conflict resolution only in the context of generating a correct partial-order plan. Conflict resolution, however, has much wider application ranges. For example, in electronics design, one has to place a set of electronic devices onto a small board area. Some of the devices might have adverse properties such as generating heat and magnetic-electric interferences. Each of these undesirable features, which can be formalized as conflicts, could be removed in a variety of ways, ranging from adding a physical shield to moving the interacting devices spatially apart. To find a consistent combination of ways for removing all conflicts requires a similar technique to what we described here.

Another example where the problem of conflict resolution often occurs is in social-political situations involving multiple social groups. This is essentially a multi-agent situation. Each group might represent a variety of objectives, many of which are often in conflict. To resolve these conflicts, one could employ a number of well-known methods, such as commissioning a go-between or offering some kind of concession.

In both these areas the conflict-resolution constraints are dramatically different from those we consider in this chapter. They are all domain-dependent in nature. However, we stress here that despite the differences in the nature of conflict-resolution constraints, at a sufficiently high level, many conflict-resolution problems could still be formalized and solved using the constraint-based technique that we have proposed in this chapter.

7.9 Summary

In this chapter, we have described a theory of conflicts and conflict-resolution methods in planning. In this theory, a conflict is modeled as a variable in a CSP the set of conflict-resolution methods is modeled as the domain of a variable. Two types of relations are described. The inconsistency relation corresponds directly to its counterpart in CSPs, and the subsumption relation provides new insights into the removal of redundancy values and variables. The formalization supports a number of efficient reasoning tasks, including arc-consistency enforcement, redundancy-removal, dead-end detection, and the ordering of conflicts in which to conduct their resolution.

In contrast to the incremental conflict-resolution methods used by a basic planning algorithm, a global conflict-resolution algorithm offers the following advantages.

Recognition of an Intelligent Ordering of Conflicts. In the presence of a large number of conflicts, the order in which the conflicts are resolved may have dramatic effects on efficiency. The global constraint-based method presented above allows the recognition of these orderings to be made. A conflict ordering heuristic can be further strengthened by recognizing "redundant" constraints, which exist because some conflict resolution constraints may be stronger than others. For example, for a set of conflicts $\mathcal{C}$, there may exist a subset $\mathcal{S}$ of $\mathcal{C}$ so that once conflicts in $\mathcal{S}$ are resolved, all other conflicts in $\mathcal{C}$ are also resolved.

Early Dead-end Detection. A plan in which conflicts cannot be resolved together corresponds to a dead end in a planner's search space. The ability to detect such dead ends early is vital to a planner's efficiency. In many situations, unresolvable conflicts can be detected with relatively low costs when considered together, but they may not be obvious when only a single conflict is considered at a time. Thus, a planning system based on the incremental method for conflict resolution may incur expensive computation due to the expansion of plans that eventually leads to dead ends.

Preprocessing of Constraint Network. Considering conflicts in a global manner permits the application of arc-consistency algorithm to the conflict set, which may lead to the discovery that some constraints can never participate in any final solutions.

7.10 Background

Conflict resolution has been a central problem in planning. Its development therefore followed a steady pattern of evolution; the reader could trace much of the historical background in many foundational planning papers. These include [142], [130], [128], [132], [146], [28], [94].

The material described in this chapter was adapted from an Artificial Intelligence journal article by Yang [154], which in turn evolved from an AAAI-90 paper that proposed to use an algebra of conflicts for planning [151]. The implementations of the COMBINE algorithm applied a method by Hertzberg and Horz for classifying conflicts [65]. Ephrati and Rosenschein considered a similar conflict-resolution problem in the context of distributed, multi-agent planning [42]. The STATIC system is described in [44], PRODIGY in [25], and Smith-Peot method in [124].

8. Plan Merging

Problem decomposition can improve planning efficiency by partitioning a complex problem into simpler parts. Since the sub-problems are solved separately, this planning strategy can also incur redundancy. To remove the redundant parts of a plan, some plan-steps can be *merged*, producing a plan that is less costly to execute.

This chapter provides a precise definition for plan merging and presents optimal and approximation algorithms for finding minimal-cost merged plans. The optimal plan-merging algorithm applies dynamic programming to find a minimal-cost merged plan from a partial-order plan. To cope with plans with large sizes, we also present an approximation algorithm along with an analysis of the quality of its output, in worst and average cases.

8.1 The Value of Plan Merging

The value of utilizing operator-overlapping interactions, or helpful interactions, was recognized early in AI planning research [114, 132, 147]. A helpful goal interaction occurs in a plan, or among several plans, if it enables a reduction in plan costs. An important type of helpful goal interaction occurs when certain operators in a plan can be grouped, or *merged*, together in such a way as to make the resulting plan more efficient to execute. This happens often in domains where redundant setup and restore operations can be eliminated in the execution of consecutive tasks and where redundant journeys can be eliminated by fetching multiple objects at once. In the related work section, we review another use of plan merging, for speeding up the reasoning process for planning, through a notion known as *overloading* [108].

In Chapter 6 we have shown a plan merging problem in the painting domain. Here we consider a story given by Wilensky [145].

> John was planning to go camping for a week. He went to the supermarket to buy a week's worth of groceries.

The main character in this example, John, had a set of subgoals to achieve, each subgoal being to buy food for a meal during the camping week. However, instead of making several shopping trips separately for each individual meal, John was able to *merge* the plans for the subgoals, and achieve them

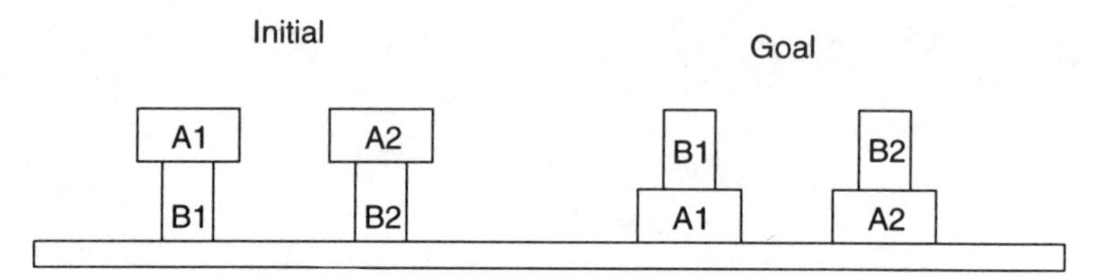

Fig. 8.1. A variable-sized blocks world

simultaneously with a single trip to the market. The resultant *merged* plan is more efficient to execute than the separate ones.

Merging actions for reducing plan costs, as demonstrated above , is typical of the kind of optimization task people do for general transportation problems. For example, suppose that cargo items A and B both need be delivered from location L_1 to location L_2 by planes. If the two delivery goals are planned separately, then two airplanes would be required for both A and B. For each item, separate loading and unloading operations are needed. However, if A and B fit into one plane, then combining the two sub-plans can produce a more efficient overall plan for both goals. This *merged plan* requires one combined loading and unloading operation for A and B, as well as a single flight operation. The result is a plan that is considerably cheaper than the original ones.

Plan merging is equally important for minimizing the costs of robot task plans. As an example, consider a blocks world problem with blocks of different sizes. To pick up one block of a certain size, the robot arm has to mount a gripper of an appropriate size. Suppose that only one robot arm exists, and in order to grab a block of a different size, the robot has to unmount the current gripper and mount the gripper with the new size. In this case, it is more efficient to group block-stacking operations that use the same type of grippers. An example is shown in Figure 8.1.

The blocks world problem is illustrative of the role of operator merging in robotic-assembly domains. Identical plan-merging issues also arise in the domain of automated manufacturing where process plans for metal-cutting [76], set-up operations [63] and tool-approach directions [93] need to be optimized. Similarly, in the area of query optimization in database systems [117], as well as domains having multiple agents [41, 111], plan merging in multiple plans seems inevitable.

In developing a computational theory on plan merging, we address the following questions:

1. What are the conditions under which plan steps in a plan can be merged?
2. What is the computational complexity of optimal plan merging?
3. What are the optimal algorithms for plan merging? When are these algorithms feasible in practice?

4. Where the optimal algorithms are infeasible to apply, what are the approximation algorithms that can perform efficiently? What is the quality of the plans produced by these algorithms in worst and average-cases?

Our approach to addressing the above questions is to combine formal methods developed in Artificial Intelligence and computational techniques from Operations Research (OR). In particular, we first propose a formalization of plan merging using the STRIPS operator definitions. Based on the formalization, we present a dynamic programming algorithm for determining the optimal solution. While the dynamic programming algorithm has traditionally been formulated for inputs corresponding to matrices of symbols, AI planning has been concerned about plans represented as partially-ordered operator sets. To bridge the two different fields, we also extend the dynamic programming method to handling partially-ordered plans in a novel way.

One drawback of the dynamic programming method is that it becomes computationally infeasible for problems of larger sizes. While we are able to phrase an optimal algorithm for general purpose, domain-independent plan merging, its run-time requirements may be prohibitive for inputs of large sizes. To make the planning problem more tractable, most existing systems that consider helpful interactions employ certain kinds of greedy algorithms for plan merging [63, 114, 132, 147]. Thus, we also describe a specification of an approximation algorithm for merging plans, and analyze the quality of its outputs in worst and average cases. The approximation algorithm has linear time-complexity in the number of plan steps of the input plan.

8.2 Formal Description

8.2.1 A Formal Definition for Plan Merging

Recall that a plan Π consists of a set of steps related to each other via a set of constraints. These steps are instantiations of a set of operator schemas. For simplicity, we first assume that all operator instantiations and steps are fully instantiated; they don't contain variable parameters.

For a given plan Π, there may be some steps in the plan that can be grouped together and replaced by a less costly step that achieves all the useful effects of the grouped steps. In such a case, we say that the steps are *mergeable.*

The notion of merging plan steps can be illustrated graphically in Figure 8.2. In this figure, there are two plans to be combined, `Plan1` and `Plan2`. The steps $S2$ and $T2$ can be merged together if there exists an operator M which can replace the collective functions of these two steps, where M is less costly than the sum of $S2$ and $T2$. M provides the functions of steps $S2$ and $T2$ if it can replace the two steps in their respective causal links.

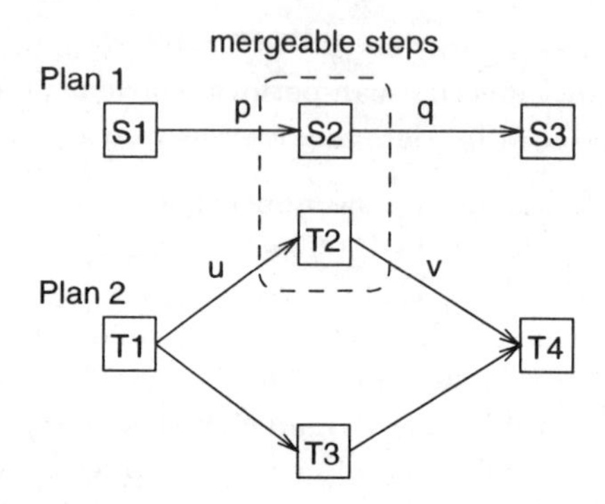

Fig. 8.2. Merging plan steps

We now make the concept of plan merging more precisely. We say that a set of steps Σ in a plan Π can be *grouped together* if no other step outside of Σ is ordered between any pair of steps in Σ. More precisely,

Definition 8.2.1 (Step Groups)
A set of steps Σ in a plan Π can be grouped together, if

$$\forall \mathbf{s}_i, \mathbf{s}_j \in \Sigma, \neg\exists \mathbf{s}_k \in (\mathit{Steps}(\Pi) - \Sigma) \text{ such that } \mathbf{s}_i \mapsto \mathbf{s}_k \mapsto \mathbf{s}_j.$$

As an example, the set of steps $\{S2, T2\}$ in Figure 8.2 can be grouped together, whereas $\{S1, T2, S3\}$ cannot.

A set of steps that can be grouped together can behave like a single step. Let Σ be a set of steps that can be grouped together. Some effects of the steps in Σ are *useful* in Π, because they establish the preconditions of some other steps that are outside of Σ. Other effects are simply side-effects of the steps in Σ and do not serve any purpose as far as the correctness of Π is concerned. We use Useful-Effects(Σ, Π) to denote the set of all useful effects of the steps in Σ. Likewise, we use Net-Preconds(Σ, Π) to denote the set of all preconditions of the steps in Σ not already achieved by any steps in Σ.

More formally,

$$\text{Useful-Effects}(\Sigma, \Pi) =$$
$$\cup_{\mathbf{s} \in \Sigma}\{e \in \text{Eff}(\mathbf{s}) \mid \exists \mathbf{s}_j \in (\text{Steps}(\Pi) - \Sigma).\ \langle \mathbf{s} \xrightarrow{e} \mathbf{s}_j \rangle \in \text{C-Links}(\Pi)\},$$
$$\text{Net-Preconds}(\Sigma, \Pi) =$$
$$\cup_{\mathbf{s} \in \Sigma}\{p \in \text{Pre}(\mathbf{s}) \mid \exists \mathbf{s}_i \in (\text{Steps}(\Pi) - \Sigma).\ \langle \mathbf{s}_i \xrightarrow{p} \mathbf{s} \rangle \in \text{C-Links}(\Pi)\}.$$

As an example, let Σ be $\{S2, T2\}$, and our plan be the one shown in Figure 8.2. Then Useful-Effects$(\Sigma, \Pi) = \{q, v\}$ and Net-Preconds$(\Sigma, \Pi) = \{p, u\}$.

Given a set Σ of steps, suppose that they can be grouped together. Σ is *mergeable* if Σ as a whole can be replaced by a cheaper step without affecting the correctness of the plan. More precisely, Σ is mergeable in Π if there exists an operator μ such that

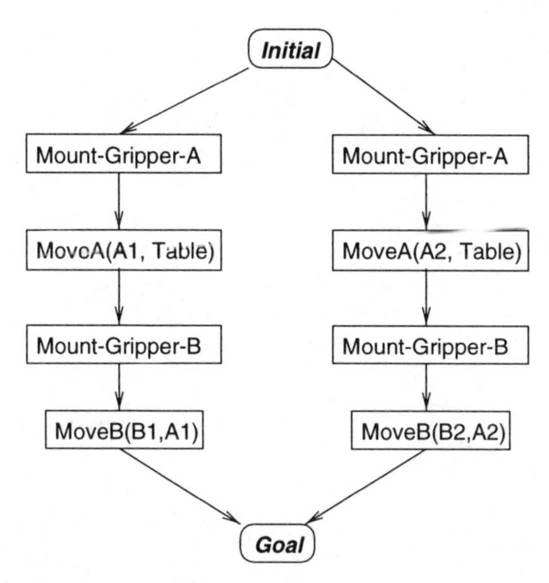

Fig. 8.3. A blocks world plan

1. Σ can be grouped together in Π,
2. Pre(μ) $\subseteq$ Net-Preconds(Σ, Π), and Useful-Effects(Σ, Π) $\subseteq$ Eff(μ). That is, the step μ can be used to achieve all the useful effects of the steps in Σ while requiring only a subset of their preconditions.
3. Cost(μ) $<$ Cost(Σ).

The step μ is called a *merged* step of Σ in the plan Π. Because a set of steps can potentially be merged and replaced by several alternative operators, each with its own cost and precondition and effect sets, we denote the set of all merged operators by $\text{MERGE}_{\Pi}(\Sigma)$ (or simply $\text{MERGE}(\Sigma)$ if it is clear about the plan Π).

8.2.2 An Example

The definition for step merging covers the examples given in the previous section. Below, we illustrate the definition through the blocks world example.

Consider again the multi-gripper blocks-world problem with blocks of different sizes (see Figure 8.1). A plan for solving the goals is shown in Figure 8.3. Suppose that `can-grab-A` is a precondition of the step `moveA`, and `can-grab-B` is a precondition of the step `moveB`. Then the following causal links establish precondition-dependencies between the steps in the plan:

$\langle \texttt{mount-gripper-A} \xrightarrow{p_1} \texttt{moveA}(A1, \textit{Table})\rangle$, where $p_1 = \texttt{can-grab-A}$
$\langle \texttt{mount-gripper-A} \xrightarrow{p_1} \texttt{moveA}(A2, \textit{Table})\rangle$,
$\langle \texttt{mount-gripper-B} \xrightarrow{p_2} \texttt{moveB}(B1, A1)\rangle$, where $p_2 = \texttt{can-grab-B}$, and
$\langle \texttt{mount-gripper-B} \xrightarrow{p_2} \texttt{moveB}(B2, A2)\rangle$,

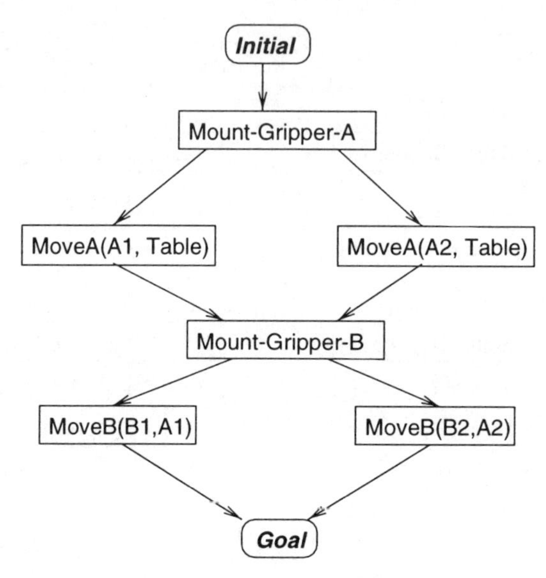

Fig. 8.4. Merged blocks world plan

The `mount-gripper-A` step in this example involves removing any previously mounted gripper and mounting a gripper of a new size. In the plan, the two `mount-gripper-A` steps can be merged into a single instance of the step. So can the `mount-gripper-B` steps. This can be verified by the following facts:

1. The two `mount-gripper-A` steps can be *grouped* together in the plan, since no other steps are necessarily between them.
2. The set Useful-Effects({`mount-gripper-A`, `mount-gripper-A`}) is exactly {`can-grab-A`}, which is identical to the effect of a single `mount-gripper-A` step. Similarly, the subset condition for net-preconditions is also satisfied.
3. Cost(`mount-gripper-A`) $<2*$ Cost(`mount-gripper-A`).

After a similar merging of the two `mount-gripper-B` steps, the plan after merging is shown in Figure 8.4.

There are several ways to extend the definition for plan merging. For example, we could allow the merged operator μ to assert only a subset of the useful effects of the steps in Σ, as long as the rest of the merged plan could *restore* the missing effects. We do not delve into the detailed case study of such a definition, but instead allow the readers to modify the definition depending on their particular problem domain. Our algorithms below could be easily adapted to these new definitions.

8.2.3 Impact on Correctness

Suppose that a set Σ of steps are merged into an operator μ. In addition to the useful effects of μ, there may also be other side effects that μ achieves.

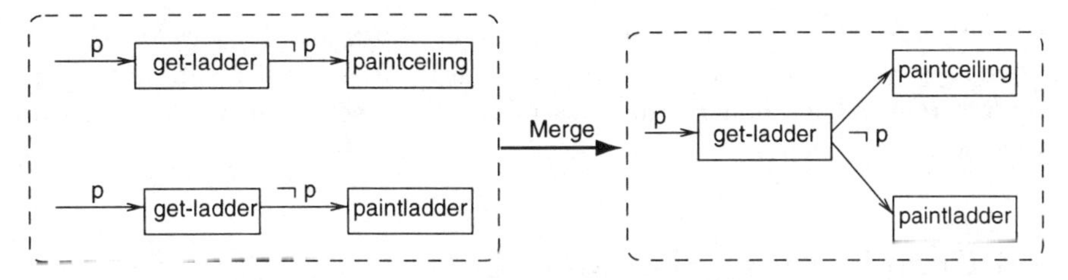

Fig. 8.5. An example where merging helps conflict resolution

Some of these side effects might be harmless, others might work against the correctness of the resulting plan by threatening an existing causal link in the plan. Thus, in deciding whether to merge a set of steps into a particular operator, cost is not the only issue; plan correctness is a factor that must also be taken into account.

To preserve plan correctness during merging, we could establish certain conditions for the merged operator to satisfy. We require that no side effects of μ negate the causal links of the resulting plan. This condition is stated formally as follows:

Condition **[Correctness Preserving].** An operator $\mu = \text{MERGE}_{\Pi}(\Sigma)$ satisfies the correctness-preserving condition if

1. For every causal link $\langle \mathbf{s}_i \xrightarrow{p} \mathbf{s}_j \rangle \in$ C-Links(Π), if μ can possibly be between $\mathbf{s}_i$ and $\mathbf{s}_j$, then for every effect e of μ that is not a member of Useful-Effects(Σ, Π), e does not unify with $\neg p$.
2. The preconditions of μ is a subset of Net-Preconds(Σ, Π).

When this condition holds and when the set Σ is replaced by μ in the plan, all preconditions of μ are already satisfied and all causal links established by members of Σ are preserved. In addition, no side effects of μ pose threats on an existing causal link. The resulting plan is therefore still correct. It can further be shown by induction that the correctness of a plan is preserved by merging steps in a plan any number of times, as long as the correctness-preserving condition is satisfied.

Plan merging can not only reduce plan costs, but also be used to resolve conflicts in a plan. As an example, consider a plan for painting both a ceiling and a door (shown in Figure 8.5). The literal p in the figure states the condition that the ladder is available. On the left-hand side of the figure, we see two plans, each with a separate `get-ladder` step conflicting with a `get-ladder` step in the other plan. The conflict arises because after getting the ladder for executing one painting plan, the agent did not specify releasing the ladder in order for the other plan to be executed. After merging the two actions, we obtain a plan on the right-hand side of the figure, where the conflict disappears.

This example shows that plan merging can help improve the correctness of a plan by reducing the number of conflicts. This additional method of conflict resolution by plan merging is important especially when plans are developed separately. It is not hard to see how we can incorporate it as an additional conflict-removal method in the constraint-based technique we discussed in Chapter 7.

8.3 Complexity of Plan Merging

Two important issues arise in plan merging. The first is deciding what steps to merge in a partial-order plan. The second is how to merge them. Below, we analyze these two problems separately.

8.3.1 Deciding What to Merge

Deciding what to merge may be computationally difficult if no additional domain knowledge is given. This is because for a given plan, it might be necessary to examine the useful effects and net preconditions of every subset of steps, and the number of such subsets can be large. To make the process more efficient, various kinds of domain knowledge can be employed. For example, one way for the steps in Σ to be merged is when they contain various sub-steps which cancel each other out, in which case the merged step μ would correspond to the set of steps in Σ with these sub-steps removed. In manufacturing domains where it is desirable to minimize set-up costs, this situation occurs often and can be profitably employed [100]. This case also corresponds to what Wilensky calls "partial-plan merging" [145]. Another case is when all the steps in Σ share a common schema, and in this case, the goals of these steps can be achieved by executing the schema only once. This case corresponds to what Wilensky calls "common-schema merging."

To make the problem sufficiently concise, we make a number of assumptions.

Operator-type Assumption. We assume that sufficient domain knowledge is available about what steps can be merged, and for each set of these steps, what the merged steps are. In particular, we assume that each operator schema in a domain description is associated with an *operator type.* For example, in the blocks world domain, a gripper-changing operation for switching to a particular gripper has a type which corresponds to the size of the gripper. As another example, in the painting domain all `get-brush` operators might be associated with the same type. Given this *typing* knowledge, we concentrate on the second issue, that of finding and analyzing methods for computing an optimal plan.

Unique Merged-operator Assumption. Given that the mergeable steps are known in a plan, one more complication still exists. For a given set Σ of steps to be merged, there may be several alternative merged steps, $\{\mu_1, \ldots, \mu_i\}$, to choose from, each with a different set of preconditions, effects and cost value. To further remove this complication, we assume that for each set Σ of mergeable steps, there is a unique merged operator μ. This restriction can be easily relaxed by slightly augmenting the dynamic programming algorithm introduced in the next section.

Mergeability Assumption. We assume that two or more plan steps can be merged only if they are unordered with each other. If they are ordered so that one is before another, then in fact the subsequent algorithms only require minimal change. However, to keep our model clear and concise, we carry through our presentation with this assumption.

8.3.2 Deciding How to Merge

Not all mergeable steps can be simultaneously merged. The partial order defined on Π steps may render some pairs of mergings inconsistent; merging one set of steps may make merging others impossible. For example, consider the two plans

$$A1 \mapsto C1 \mapsto B1 \text{ and } A2 \mapsto B2 \mapsto C2,$$

where the plan steps can have three types, A, B and C. Thus, after step merging, one merged plan is

$$\text{MERGE}(\{A1, A2\}) \mapsto B2 \mapsto \text{MERGE}(\{C1, C2\}) \mapsto B1,$$

while another merged plan is

$$\text{MERGE}(\{A1, A2\}) \mapsto C1 \mapsto \text{MERGE}(\{B1, B2\}) \mapsto C2.$$

However, it is impossible to merge both pairs $B1, B2$ and $C1, C2$ without violating the partial order (see Figure 8.6).

We now consider the computational complexity as a result of the above complication. The problem is to decide which set of mergeable steps to merge, when a partial order prevents all of them from being merged simultaneously. We show that the problem of finding an optimally merged plan is NP-hard.

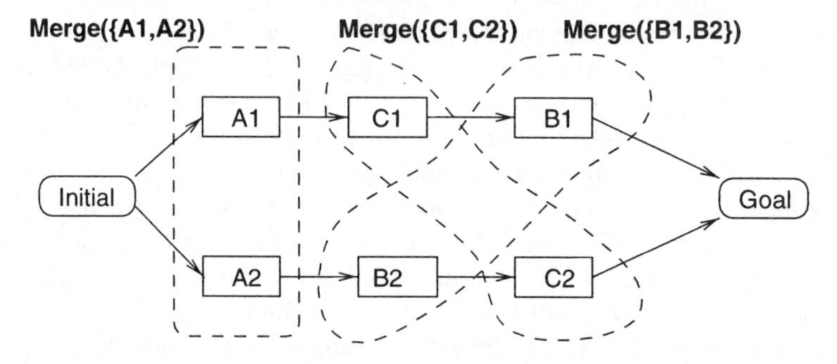

Fig. 8.6. Computational complexity of plan merging

We first introduce the *Shortest Common Supersequence (SCS)* problem: given m sequences of characters $S_1, S_2, \ldots, S_m$, what is the shortest sequence S such that each S_i is a subsequence of S? This problem has been shown to be NP-hard [55].

We give a transformation of a SCS problem to a plan-merging problem. In a SCS problem, for each sequence $S_i = (c_{i1}, c_{i2}, \ldots, c_{ik_i})$, we define a plan $P_i = (a_{i1}, a_{i2}, \ldots, a_{ik_i})$, where all steps have the same costs. We also define a goal G_i achieved by P_i. The only kind of interaction that occurs is an operator-overlapping interaction; we define two steps a_{ij} and a_{kl} to be mergeable if and only if $c_{ij} = c_{kl}$. It follows that the plans $P_1, \ldots, P_m$ can be merged into a plan $P = (a_1, a_2, \ldots, a_k)$ if and only if the sequence $S = (c_1, c_2, \ldots, c_k)$ is a supersequence for $S_1, \ldots, S_m$.

One way to get around the NP-hardness of a computational problem is to set an input parameter constant. In many plan-merging problems this strategy could be applied by setting the total number of goals to be achieved *constant*. Thus, for these naturally occurring cases the number of input plans is also constant. In such cases, the optimal solution to the SCS problem can still be found in polynomial time using a dynamic programming algorithm.

8.4 Optimal Plan Merging

We now describe how to find an optimally merged plan using *dynamic programming*. A set of partially ordered plans can be combined into a single plan Π, the steps of which correspond to a network of nodes, where each node is a plan step, and an edge between two nodes is a precedence relation between the corresponding steps. Figure 8.7 shows an example of viewing a three-plan merging problem as a network problem.

The intuition behind dynamic programming is inductive computation. One starts from a small subset of the plan steps and computes an optimal merged plan for the subset. Then one can extend the subset by including more steps, and compute the cost of the merged plan for the enlarged subset, based on both the cost of each smaller subset, and the costs of the new plan steps. For an introduction to dynamic programming, see Chapter 4.

We consider the process of computation as one of a left-to-right push of a *frontier* which divides the plan into two parts. The part on the left denotes a subset of the plan steps for which an optimal merged plan has already been found, and the one to the right those plan steps that are yet to be included. The frontier includes all nodes (or plan steps) that are unordered with respect to each other in the plan, and that are being considered for merging (see Figure 8.7). Let F be the set of nodes on the frontier. Let Σ_i be the set of *all* nodes on the frontier F and whose corresponding plan steps have the same type, Type$(i), i = 1, 2, \ldots, m$. These are nodes that can be merged together on F. Finally, let σ_i be a subset of $\Sigma_i, i = 1, 2, \ldots m$. Then if Optimal$(F)$ denotes the cost of the optimal merged plan for the subset of

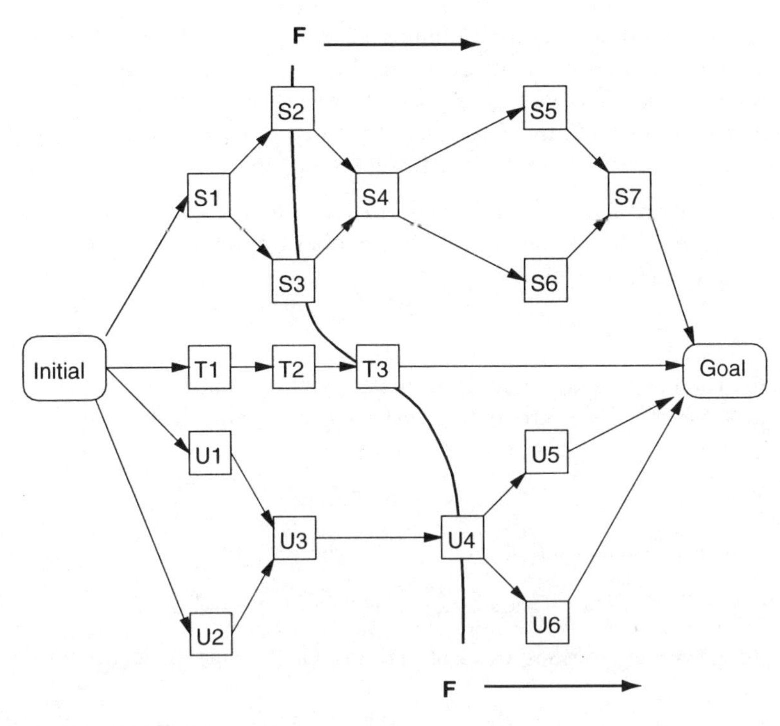

Fig. 8.7. Intuition of dynamic programming

plan steps upto and including the frontier F, the cost can be formulated as follows:

$$\text{Optimal}(F) = \min \begin{cases} \text{Optimal}(F - \sigma_1) + \text{Cost}(\mu_1), & \forall \sigma_1 \subseteq \Sigma_1 \\ \text{Optimal}(F - \sigma_2) + \text{Cost}(\mu_2), & \forall \sigma_2 \subseteq \Sigma_2 \\ \dots & \dots \\ \text{Optimal}(F - \sigma_m) + \text{Cost}(\mu_m), & \forall \sigma_m \subseteq \Sigma_m \end{cases} \quad (8.1)$$

In the above formula, μ_i is a merged operator for set σ_i, $i = 1, 2, \dots, m$ respectively.

Having designed a method for computing the cost of the optimal merged plan, we can then generalize the computation so that along each frontier, the optimal choice in step-merging, σ, is remembered. The sequence of merged steps, as we push the frontier F from left to right, will give rise to a set of plan steps that should be merged to result in an optimal merged plan.

More specifically, for each frontier F, let σ be a set of merged steps on F which gives rise to a minimal value as computed by equation (8.1). Let Merged-Steps(F) be the remembered subsets of plan steps up to and including the frontier F that should be merged. We have

$$\text{Merged-Steps}(F) = \text{Merged-Steps}(F - \sigma) \cup \{\sigma\}$$

To apply the dynamic programming algorithm, DYNAMIC, of Chapter 4, we must specify a parent-child relationship between states. In our formulation, each frontier F is a state. A parent of F is a frontier E such that the "difference" between E and F is a set of plan steps σ to be merged. Let σ be a subset of F. Let β be a set of plan steps such that

- every member of β is *immediately ordered before* a member of σ. A plan step $\mathbf{s}_i$ is immediately before $\mathbf{s}_j$ if in the plan there are no other steps that are ordered between $\mathbf{s}_i$ and $\mathbf{s}_j$. And
- every member of β is *unordered* with every other member of β and every member of $F - \sigma$.

Consider the largest such β, and call it Patch(σ). Patch(σ) can be considered as a patch for the "hole" created by cutting a set σ from the frontier F. Given the set Patch(σ), a parent state of F is

$$P = (F - \sigma) \cup \text{Patch}(\sigma)$$

The set of all parents of F, Parents(F), is the set

$$\text{Parents}(F) = \{(F - \sigma) \cup \text{Patch}(\sigma) \mid \forall \sigma \in \Sigma_i,\ i = 1, 2, \ldots, m\} \qquad (8.2)$$

This set represents various ways of arriving at the current frontier from all previous ones.

Finally, let $d(P, F)$ be a decision in the dynamic programming algorithm (see Chapter 4, Table 4.4), for going from a parent state E to a child state F, via merging of a set σ of plan steps. This is denoted as MERGE(σ). Having these functions in place, a dynamic programming implementation of optimal plan merging is simply a function call DYNAMIC($\{\mathbf{s}_{finish}\}$). We will refer to this instantiation of the dynamic programming algorithm for optimal plan merging as OPTIMALMERGE(Π).

The dynamic programming algorithm is listed in Table 8.1.

Now we consider the complexity of the algorithm. Let k be the maximum size of a frontier; k determines how "fat" the partial-order plan is. If the plans are all linear sequences of steps, then k is the number of goals to be achieved, which we represented by m in Chapter 6.

To carry out the computation for equation (8.1), it is necessary to enumerate all subsets of the mergeable steps on a given frontier. This takes time $O(2^k)$. Then the OPTIMALMERGE algorithm iterates through all frontiers in the plan, the total number of which is l^k where l is the maximal length of Π. Here the *length* of a partially ordered plan is defined as the maximum number of steps that form a linear chain from start of the plan to finish. Thus, the total amount of time taken by the algorithm is

$$\text{Time(OPTIMALMERGE)} = O(2^k l^k)$$

This formula shows that when the width of a plan Π is fixed at a constant, the time complexity of the OPTIMALMERGE algorithm increases polynomially

Table 8.1. The OPTIMALMERGE algorithm

Algorithm OPTIMALMERGE(Π)
Input: A plan Π.
Output: An optimal merged plan.

```
1.  call UPTOFRONTIER({s_finish});
2.  for each set σ in Opt-Decisions({s_finish}) do
3.      merge σ in plan Π;
4.  end for
5.  return(Π);
```

Subroutine UPTOFRONTIER(F)
Input: A search frontier F.
Output: The cost of an optimal merged plan up to the frontier. A global variable Opt-Decisions(F) records all steps that should be merged.

```
1.  if F is the start step s_init then
2.      return(0);
3.  else
4.      Parent-List := Parents(F), calculated using Equation 8.2;
5.      let S be a parent of F in Parent-List such that
            Cost := (Cost(MERGE(σ)) + UPTOFRONTIER(S)) is minimal;
6.      Opt-Decisions(F) := Opt-Decisions(S)∪{σ};
7.  end if;
8.  return(Cost);
```

---

with the length of plans. This indicates that the algorithm is most useful when the plan is a small collection of almost totally ordered plans. In fact, this dynamic programming algorithm is the basis for merging two sequences of text strings in UNIX systems using "diff." However, when k becomes large, the computation can become very expensive.

8.5 Approximate Plan Merging

For problems of large sizes, the complexity of the dynamic programming methods may be too high to be practical. An alternative choice is to use approximation algorithms that operate in low-order polynomial time but output a suboptimal merged plan.

In the past, such planning systems as NOAH [114], SIPE [147], and MACHINIST [63] have resorted to the application of greedy algorithms for plan merging. However, the approximation algorithms used by the systems were never clearly specified, nor was it known how these algorithms perform in the worst and average cases. In this section, we consider a greedy algorithm that is intuitively appealing, and present complexity analysis on its performance.

8.5.1 Algorithm ApproxMerge

Recall that the dynamic programming algorithm OptimalMerge compares all possible merges on all frontiers in the plan. This is in fact one source of inefficiency. In the approximation algorithm, we simplify the OptimalMerge algorithm by

- considering only a limited number of frontiers; and
- considering only a limited number of sets of steps for merging, along a frontier.

As with OptimalMerge we assume that our plans are arranged in a left to right order, and we push a frontier, Start(Π), in that direction. Once a set of steps is decided to be merged on a frontier, we take that merging operation as *final*, and take the steps off the plan. This forms a new frontier. In this way, every step is considered by the approximation algorithm *only once*, and thus the time complexity of the algorithm is *linear* in the total number of steps in Π.

To make the above picture more precise, we formally define a frontier of a plan Π as Start(Π). Given a plan Π, one can find a subset of steps with no other preceding steps. Formally,

$$\text{Start}(\Pi) = \{\mathbf{s} \in \text{Steps}(\Pi) \mid \neg\exists \mathbf{s}_i \in \text{Steps}(\Pi).\ \mathbf{s}_i \mapsto \mathbf{s}\}$$

In addition, let Σ be a set of plan steps in Π. The function $\texttt{Remove}(\Sigma, \Pi)$ returns the plan Π with all steps in Σ removed.

The approximation algorithm is shown in Table 8.2. It classifies all steps on the current frontier as denoted by Start(Π) by operator types. It then merges all steps in a subclass Σ_i with maximal cardinality.

8.5.2 Example

We trace the ApproxMerge algorithm on the plan in Figure 8.6. For simplicity, we assume that the cost of a set of steps is the cardinality of that set. Recall that in this example, any steps with the same alphabet can be merged.

Initially, Start(Π) = $\{\mathbf{s}_{init}\}$. After the start step is removed from the plan, the current frontier contains $\Sigma_A = \{A_1,\ A_2\}$. These two steps can be merged, giving rise to a current record of mergeable steps Merged-Steps = $\{\{A_1,\ A_2\}\}$.

After Σ_A is removed from the plan (Step 5 of the algorithm), the current frontier consists of $\{C_1, B_2\}$. An arbitrary choice selects $\Sigma_C = \{C_1\}$. After

Table 8.2. The ApproxMerge algorithm

Algorithm ApproxMerge(Π)
Input: A plan Π.
External Function: ApplyMerge(S, Π) returns Π with the steps in S merged.
Output: A merged plan.

1. Merged-Steps := $\emptyset$; Π' := Π;
   while $\Pi \neq \emptyset$ **do**
2. partition Start(Π) into m subsets Σ_i by type;
3. let Σ_j be a set such that
   (Cost(Σ_j) – Cost(Merge(Σ_j))) is maximal;
4. Merged-Steps := Merged-Steps$\cup\{\Sigma_j\}$;
5. Π := `Remove`(Σ_j, Π);
6. **end while**
7. **return**(ApplyMerge(Merged-Steps, Π'));

Σ_C is removed from the plan, Start(Π) $= \{B_1, B_2\}$, which can be merged into a B step. After the last step C_2 is removed from the plan, only the finish step remains, and the algorithm terminates. The merged plan is

$$\text{Merge}(\{A_1, A_2\})\mapsto C_1 \mapsto \text{Merge}(\{B_1, B_2\})\mapsto C_2$$

8.5.3 Complexity

Since Algorithm ApproxMerge considers each step only once, its time complexity is linear in the number of steps. We are hence more interested in the *quality* of the plan it produces in worst and average cases.

For ease of analysis of the worst case complexity in plan quality, we assume that the plan Π consists of a set of totally ordered plans. Also for simplicity we assume that the *cost of a plan* is the number of steps (excluding the start and finish steps) in the plan.

Theorem 8.5.1 (Worst Case). *Given a set of n totally ordered plans, each of length n, the worst-case cost of* ApproxMerge *is* $\Theta(n \log n)$.

Proof. We first give a set of n plans each containing n steps of type in $\{\alpha, \beta\}$, from which algorithm ApproxMerge will produce a superplan of size $\Omega(n \log n)$. For operator of $u > 2$ types, the proof is similar. We conveniently arrange the plans into an n by n matrix M such that each row of M corresponds to a plan, one step per entry. We recursively construct M: Its first $\lfloor n/2 \rfloor + 1$ rows are all α. The first column of the rest of the rows contains β. Therefore, according to its description, algorithm ApproxMerge always chooses the first column of the first $\lfloor n/2 \rfloor + 1$ rows to merge at the first n steps. At step $n+1$, ApproxMerge chooses the first column of β from the rest of rows. After the execution of the first $n+1$ steps, we are left with a matrix of roughly $\lfloor n/2 \rfloor - 1$ by $n-1$. If we recursively apply above construction to the remaining matrix, it is apparent that the merge process will go on for $\Omega(n \log n)$ steps.

We now show that ApproxMerge always produces a super-plan no more than size $O(n \log n)$ on plans drawn from a binary type for plan steps (for fixed number of operator types $u > 2$ the proof is similar). The minimum number of steps possibly merged by ApproxMerge in successive steps is a monotonically decreasing function $f(k)$ of step number k. The worst case behavior of ApproxMerge maximizes the step number at which $f(k)$ vanishes. (We approximate $f(k)$ by its continuous analog $f(x)$.) All realizations of $f(x)$ have $\int_0^\infty f(x)dx = n^2$, that is, ApproxMerge merges all n^2 steps into a super-plan. As well, $f(x) \geq (2n)^{-1} \int_x^\infty f(y)dy$, for ApproxMerge is able to merge at least half of the number of plans not yet exhausted by step k. The worst case behavior of ApproxMerge is obtained by setting $f(x)$ to its minimum value for all x. In this case, $f(0) = n/2$ and $f(x) = (2n)^{-1} \int_x^\infty f(y)dy$, with solution $f(x) = (n/2)e^{-x/2n}$. For large n, this function first vanishes when $x = cn \log n$ for some $c > 0$. □

We now examine the average case cost of ApproxMerge. We assume that the average case is one in which every operator is equally likely to appear in a plan. Under this uniform-distribution assumption, we can employ a probabilistic analysis of the algorithm. The proof, which we omit here, is a bit involved. Interested readers could consult [49]. We provide the final conclusion in the theorem below.

Theorem 8.5.2 (Average Case). *Given n randomly generated, totally ordered plans, each containing at most n steps taken from an operator set $\{\alpha_1, \alpha_2, \ldots, \alpha_u\}$, and for any small positive constant ϵ, the average case cost of* ApproxMerge *is no greater than $n(u+1)/2 + O(n^{1/2+\epsilon} \log n)$, where u is the total number of mergeable operator types.*

The proof of this theorem assumes that the plan sizes n are asymptotically large. This theorem established that the linear-time algorithm ApproxMerge is in fact expected to perform well when mergeable step types are uniformly distributed over a plan and when the plan sizes are large.

8.6 Related Work

8.6.1 Critics in Noah, Sipe, and Nonlin

Sacerdoti's Noah[114] system is one of the first planners to explicitly seek opportunities for plan merging. It relies on its set of critics to handle possible interactions among the different parts of a plan. Three critics are introduced for the purpose of improving the quality of plans, including "eliminate redundant preconditions," "use existing objects" and "optimize disjuncts." The "eliminate redundant preconditions" critic can be considered as merging two or more steps that are used to achieve the same preconditions, as long as no precedence relation in the plan is violated. Wilkins' SIPE [147] and Tate's NONLIN [132] are other planning systems that have relatively more advanced capabilities in step merging. For example, they are able to recognize that a goal is achievable by an existing step in the current plan. In such situations, they will impose constraints on orderings and variable bindings so that the step is used to achieve it. This process, known as *phantomization* of a goal, is also used in Kambhampati and Hendler's plan-reuse framework for reducing plan cost [73]. At a meta-planning level, Wilensky [145] considers different types of *positive goal relationships*, in the context of cognitive modeling of human problem solving.

Despite early recognition of its importance, none of the previous work has actually considered a formal, precise characterization of plan merging. Nor are algorithms for optimal and approximate plan-merging presented or analyzed. Our formalization and computational framework for plan merging could be considered as a formal basis for these systems.

8.6.2 MACHINIST

In contrast to the above domain-independent approaches to planning, Hayes [63] proposed a *domain-dependent* method for plan merging in machining domains. Through careful cognitive studies in the manufacturing domain, Hayes observed that many expert machinists often look for opportunities for machining-operator overlap during a plan-generation process. In this domain, each machining operator is described by an orientation of the part, fixtures for the part (such as clamping), and tool types (drilling, milling, etc.). Each machining feature, a hole or a pocket, could be considered as a goal, and a sequence of machining operations a plan. Because different features on the same part might involve the same fixture and tools, and because some of these identical fixture and tool operations could considerably shorten a plan, an expert machinist will not only be on the lookout for possible operator-overlaps, but also plan for these overlaps to happen.

Her observation is implemented in a rule-based system called MACHINIST. In the system an approximation algorithm is implemented for performing the merging operation. The results are compared with that by human machinists in terms of the quality of the plans produced, and the amount of time it takes to develop a plan. It was shown that MACHINIST can often out-perform an experienced human process-planner. For example, the quality of a plan produced by MACHINIST was shown to be much better than one by an expert machinist with three years of experence, and only slightly worse than one by a machinist with five years of experience.

8.6.3 Operator Overloading

Many of the formalization and computational considerations in this chapter are centered around a single notion: to make a plan less costly to execute. Plan merging, in fact, has been seriously considered by Pollack [108] and others as a machanism for speeding up the plan-reasoning process, through a notion known as *overloading*.

In [108], Pollack presents an enlightening example to illustrate the overloading concept: consider a plan for getting flour for baking a cake. Suppose that, at the moment, the agent has a number of active plans, already developed for some other goals such as to go to a bank. By recognizing the opportunities for *merging* the plans for flour-shopping and visiting the bank, the agent could choose to *overload* the banking plan to accomplish both goals. An end effect is reduced planning time for solving all goals.

As Pollack pointed out, the concept of overloading is supported by some strong cognitive evidence [64]. To implement overloading requires that we modify our problem-decomposition framework. Instead of generating alternative plans for each goal *before* merging them, one could interleave plan generation and plan merging, in order to take immediate advantage of overloading a plan. In this way, the plan found for all goals may not be the absolutely optimal one, but for a rational agent, as Simon argues in [122], the *satisficing* approach might very well be the most effective way for dealing with limited computational resources.

8.7 Open Problems

8.7.1 Order-Dependent Cost Functions

One observation by Hayes in the manufacturing domain is that the cost of a plan is often dependent on the order in which the steps in a partial-order plan are sequenced. This presents a new problem in plan merging, as in the entire chapter we have assumed that the cost of a plan is the sum of all step-costs, independent of the order. When knowledge is available on how the cost of a plan changes with step-ordering, both the dynamic programming method for optimal plan merging and the approximate plan merging method could be adapted for merging plans.

8.7.2 Hierarchical Plan Merging

Real-world problems are inherently hierarchical. Often one could separate the most important aspects of a complex problem from the rest. One criterion for such separation is cost. In a machining domain, Britanik and Marefat (see [22]) noticed that operations on a machined part could sometimes be partitioned into classes, one for fixturing, another for tooling. The fixture-setup operations usually take up so much time that these operations could be placed at a higher level of abstraction. With this hierarchical knowledge, they proposed an algorithm for merging plans in a hierarchical manner: plan merging for the fixture-setup operations is done first, that for tooling next, and so on. It is shown in [22] that this approach could dramatically reduce the time needed for performing plan-merging, while not sacrificing much of the plan quality. It would be interesting to see how this approach of hierarchical plan merging could fit under a similar formal framework for plan merging, and to study under the framework the benefits of the approach.

8.7.3 Enhancing Conflict Resolution

In this chapter we have also shown how plan-merging could be used as a method for conflict resolution. Although we have exposed the intuition through an example, we have not developed a formal model to incorporate this technique. We have instead taken the conceptually clearer approach of separately discussing plan merging and conflict resolution. It would be worthwhile to see how plan-merging and conflict resolution could be fruitfully combined in a seamless manner.

8.8 Summary

Opportunities for plan merging are particularly abundant when each plan is generated separately from the rest. In this chapter we have presented a formal model for plan merging, showing the precise meaning of merging steps to reduce plan costs, and demonstrating when such operations will not affect the correctness of a plan. We have also provided an optimal algorithm and an approximation algorithm for computing a merged plan.

8.9 Background

Most of the background material has been covered in Section 8.6. The work presented in this chapter is adapted from three sources: Yang's PhD dissertation [150], an Artificial Intelligence journal article by Foulser, Li, and Yang [49], and part of a Computational Intelligence journal article by Yang, Nau, and Hendler [158]. The dynamic programming algorithm is made considerably more concise than the one presented in [49].

9. Multiple-Goal Plan Selection

In this chapter we deal with the problem of selecting plans from a set of pre-existing alternatives. We consider two closely related versions. The first problem occurs when plans are selected based on a set of consistency criteria. The second problem arises when plans for different goals can be merged to improve efficiency.

9.1 Consistency-Based Plan Selection

So far we have discussed how to resolve conflicts among separately generated plans using constraint satisfaction. In many real-world situations, planning problems can be decomposed in such a way that for each sub-problem, there is more than one plan. Thus, in addition to the problem of combining the solutions, there is also a problem in *selecting* one plan per sub-problem, in order for them to be combined. In this section, we describe another constraint-based method for multiple-goal plan selection.

9.1.1 The Multiple-Goal Plan-Selection Problem

The *multiple-goal plan-selection problem* occurs in several practical planning domains. One example is database query-plan generation. Typically in the database domain, a query could be answered by alternative query plans, each being a sequence of database access operations. To answer multiple queries, an intelligent database-management system selects one query plan from each set and combines them together.

We demonstrate the multiple-goal plan-selection problem via a version of the painting domain. Suppose our tasks are to paint a ceiling, a ladder, and a door. Suppose also that for each goal, we have multiple alternative plans to choose from. In particular, we have

$goal_1$ = Painted(Ceiling):

Π_{11} = `machine-sand`(Ceiling)$\mapsto$`spray-paint`(Ceiling),

Π_{12} = `hand-sand`(Ceiling)$\mapsto$`spray-paint`(Ceiling),

Π_{13} = `white-paint`(Ceiling)$\mapsto$`hand-paint`(Ceiling).

$goal_2$ = Painted(Ladder):
Π_{21} = machine-sand(Ladder)$\mapsto$spray-paint(Ladder),
Π_{22} = hand-sand(Ladder)$\mapsto$spray-paint(Ladder),
Π_{23} = white-paint(Ladder)$\mapsto$hand-paint(Ladder).
$goal_3$ = Painted(Door):
Π_{31} = machine-sand(Door)$\mapsto$spray-paint(Door),
Π_{32} = hand-sand(Door)$\mapsto$spray-paint(Door),
Π_{33} = white-paint(Door)$\mapsto$hand-paint(Door).

Suppose that not all alternative plans can be consistently combined; some might be incompatible. For example, it might be the case that Π_{11} and Π_{23} cannot be used in the same global plan due to a certain side effect of spray painting. In such cases, choosing which plan to use for each subgoal is a non-trivial task.

9.1.2 A CSP Representation

We now formalize the problem. Suppose that we are given a set of decomposed goals $g_i, i = 1, 2, \ldots, n$, and a set of alternative plans $\{\Pi_{ij}, j = 1, 2, \ldots m\}$, one set for each goal g_i. Our problem is to find one plan for each goal such that all selected plans can be combined into a correct global plan.

We model this problem as a constraint satisfaction problem. In particular, let every goal set g_i be represented as a variable, every alternative plan Π_{ij} as a value in the domain of g_i. A consistency relation Consistent$_{\Pi}$ is defined between two values $A = \Pi_{ij}$ and $B = \Pi_{lk}$ such that

$$\text{Consistent}_{\Pi}(A, B) = \text{TRUE iff } \textsc{Combine}(\{\Pi_{ij}, \Pi_{lk}\}) \neq \text{Fail}$$

where COMBINE() is a procedure (Table 7.2) of resolving conflicts in a plan. See Chapter 7 for more descriptions.

In general, the relations between the variables in this CSP are not just binary; three or more values can be related to each other via a consistency relation. This feature calls for a general backtrack-based method where in every step of search, all sub-plans accumulated so far are checked against a global consistency relation. This would ensure that the final global plan is conflict-free. Nevertheless, using the binary consistency relations defined above we could prune away many incompatible values before the backtrack-based method is used.

9.1.3 A Constraint-Based Solution

As with many other constraint-satisfaction problems, our solution to the multiple-plan selection problem can be formulated in two phases, local-consistency methods and intelligent backtrack-based methods. Our adaptation of these methods is implemented in a planning system known as WATPLAN.

WATPLAN first applies local-consistency methods in an attempt to propagate consistency constraints among subsets of plans. Arc-consistency is applied to pairs of goals. When a plan Π_{ij} is found to be inconsistent with all plans for another goal, Π_{ij} can be pruned away. Similarly, we could define a *subsumption relation* between plans, such that a plan Π_{ij} subsumes Π_{uv} if the former could solve the goal , g_u, of the latter. With subsumption relations so defined, we could then apply the Var-Subsumption and Value-Subsumption Theorems defined in Chapter 7 to prune away redundant goals and plans.

Freuder's theorem (Theorem 4.5.1) also opens up an interesting opportunity for efficiently solving the multi-goal plan-selection problem. For problem-decomposition, one might discover that the interaction between the goals defines a tree-shaped network. If one can further ensure arc-consistency by studying pairs of frequently used plans for each goal-set g_i, then by Theorem 4.5.1, we know that a backtrack-free order of goals can be followed to form a global, consistent plan for all goals.

In general, however, we may not be so lucky as to have a tree-shaped decomposition of the domain. In such cases, once arc-consistency is processed, we can then use a backtrack-based method, such as forward-checking, to compose a global solution. The resulting algorithm WATPLAN is shown in Table 9.1.

The core of WATPLAN is a forward-checking algorithm, with a slight modification. As described in Chapter 4, forward-checking performs a backtrack-

Table 9.1. The WATPLAN algorithm

Algorithm WATPLAN($\mathcal{G}$, $\mathcal{P}$)
Input: A set of decomposed goals $\mathcal{G} = \{g_i, i = 1, 2, \ldots\}$, each goal is associated with a set of alternative plans $\mathcal{P} = \{P_i\}$, where each $P_i = \{\Pi_{ij}\}$.
External Functions:

- FWDCHECKING(CSP) is the forward-checking algorithm, described in Chapter 4.
- PARTIALARC(CSP) returns a CSP as a result of a partial arc-consistency processing.

Output: A global, consistent plan Solution for solving all the goals, if one exists. Otherwise Fail.

1. represent the problem $\langle \mathcal{G}, \mathcal{P} \rangle$ as a CSP;
2. let the constraint satisfaction problem be CSP;
3. CSP := PARTIALARC(CSP);
4. **if** a variable in CSP becomes empty,
5.   **then return**(Fail);
6. **end if**
7. Solution := FWDCHECKING(CSP);
8. **return**(Solution);

---

ing search with partial look-ahead. During this process, a binary consistency relation between any pair of plans could be verified between the current variable assignment and future ones. Two plans Π_1 and Π_2 are consistent if COMBINE($\{\Pi_1, \Pi_2\}$) does not fail, where the COMBINE algorithm, defined in Section 7.6, performs conflict resolution for the two plans. Once a complete assignment is accumulated, with one value for every variable, WATPLAN then checks whether *all* plans can be consistently combined, using the COMBINE algorithm. If not, backtrack occurs.

9.1.4 Evaluation—a Blocks-World Example

As predicted before, a global analysis of conflicts is expected to be particularly useful for problem solvers that are based on a problem-decomposition strategy. A situation in which problem-decomposition can be profitably applied is where there is enough domain-dependent knowledge for generating solutions to each individual sub-problem, but the conflicts among the sub-solutions need be resolved when a global solution is formed. WATPLAN is extended to interface with a set of specialists to facilitate this way of problem solving. In particular, it is assumed that an ordered set of alternative solutions has been generated by each specialist within the problem domain. WATPLAN then conducts a systematic selection of the sub-solutions, and applies its conflict-detection, preprocessing, variable-ordering and backtracking algorithms to combine the solutions. If the resultant plan can be made correct, then one such correct plan is returned. Otherwise, it returns to the previous step to select the next set of plans for combination. The process repeats until either there is a successful combination, or there is no new combination to be considered.

To test the efficiency of this strategy, experiments were done in the blocks world domain, where a planning problem is defined by the initial and final configurations of stacks of blocks on a table. The restrictions are that only one block can be moved at a time, and that a block cannot simultaneously support more than one block. The operators in this domain are $\texttt{move}(x, y, z)$, for moving a block x from y to a block z, and $\texttt{newtower}(x, y)$ for moving a block x from the top of y to the table.

One way to decompose the blocks world domain is to consider the movement of each block as the task of a specialist. Suppose that a specialist knows exactly how a block x can be moved, in the following manner: From any initial situation $\text{On}(x, y)$ to a goal situation $\text{On}(x, z)$, every block x is moved in precisely one of the following ways:

1. If $y = z =$ Table, then return $\{(\texttt{donothing})\}$ as the only solution for moving block x. `donothing` denotes an empty sub-plan, in which all goal conditions are established by the initial situation.
2. If $y =$ Table but $z \neq$ Table, then return $\{(\texttt{move}(x, \text{Table}, z))\}$.
3. If $y \neq$ Table and $z =$ Table, then return $\{(\texttt{newtower}(x, y))\}$.

4. If $y = z \neq$ Table, then return a set of two alternative solutions: $\{(\texttt{donothing}), (\texttt{newtower}(x, y) \mapsto \texttt{move}(x, \text{Table}, z))\}$.
5. Otherwise, return $\{(\texttt{move}(x, y, z)), (\texttt{newtower}(x, y) \mapsto \texttt{move}(x, \text{Table}, z))\}$.

The above domain-dependent enumeration of the movement of a block completely characterizes its possible movement for any given initial and final situations. Therefore, given a blocks-world problem, if there is a plan for all blocks, then a sub-plan exists for each individual block that can be combined to result in a correct one. In other words, WATPLAN is complete for this domain. The difficulty lies in the selection of sub-plans which can be combined to result in a final solution. When more than one block exists, a choice made may not only affect the successful movement of one block, but also make the other blocks' movements either easier or harder, or even impossible. We illustrate the selection process through the following example.
The initial situation is

On(C,A),On(A,B),On(B,Table),Clear(C),Clear(Table)

and the goal is

On(A,B),On(B,C),On(C,Table).

The initial sub-plans, which are provided by the specialists for the blocks, are listed below.

For `block A`: $\{(\texttt{donothing}), (\texttt{newtower}(\text{A}, \text{B}) \mapsto \texttt{move}(\text{A}, \text{Table}, \text{B}))\}$
For `block B`: $\{(\texttt{move}(\text{B}, \text{Table}, \text{C})\}$
For `block C`: $\{(\texttt{newtower}(\text{C}, \text{A}))\}$.

The first choice for sub-plan combination includes the sub-plan `donothing` for block A. However, when the three sub-plans are combined, no operator can be found in the plan that establishes the precondition Clear(B) of `move`(B,Table,C). But the second choice for combination, listed below, can be successfully merged.

For `block A` : $(\texttt{newtower}(\text{A}, \text{B}) \mapsto \texttt{move}(\text{A}, \text{Table}, \text{B}))$
For `block B` : $\texttt{move}(\text{B}, \text{Table}, \text{C})$
For `block C` : $\texttt{newtower}(\text{C}, \text{A})$

In particular, when the three sub-plans are combined, the newly-found causal links for precondition Clear(A) of `move`(A,Table,B) and precondition Clear(B) of `move`(B,Table,C) require the imposition of ordering constraints

$$\texttt{newtower}(\text{C}, \text{A}) \mapsto \texttt{move}(\text{A}, \text{Table}, \text{B}),\ \texttt{newtower}(\text{A}, \text{B}) \mapsto \texttt{move}(\text{B}, \text{Table}, \text{C}).$$

Furthermore, the following conflicts are detected:

1. `move`(A,Table,B) is a clobberer for the causal link $\langle \mathbf{s}_{init} \xrightarrow{p} \texttt{move}(B, Table, C)\rangle$ where $p =$ Clear(B), and

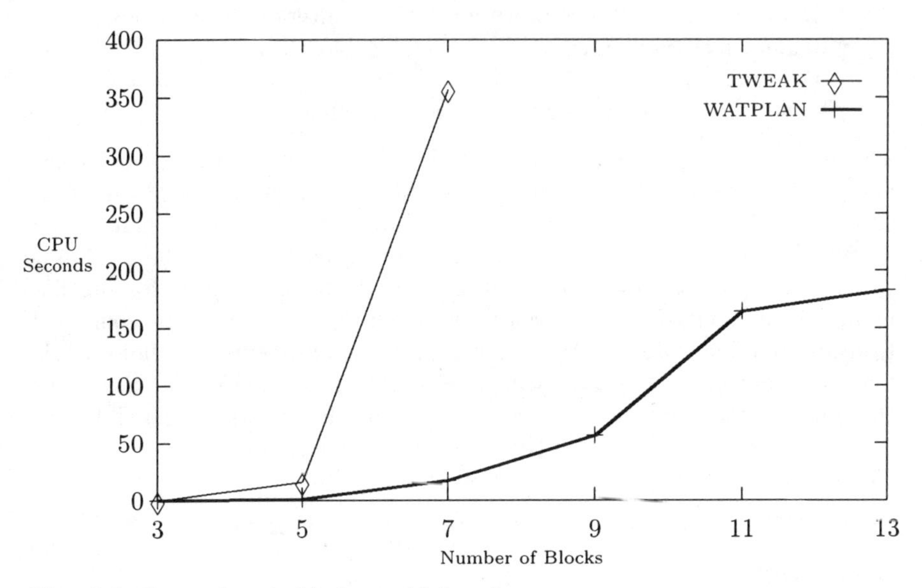

Fig. 9.1. Comparison in blocks-world domain

2. move(B,Table,C) is a clobberer for the causal link $\langle s_{init} \xrightarrow{q} \texttt{newtower}(C, A)\rangle$ where q = Clear(C).

A linear plan is obtained by resolving both conflicts:

newtower(C,A) $\mapsto$ newtower(A,B) $\mapsto$ move(B,Table,C) $\mapsto$ move(A,Table,B).

Empirical tests have been conducted with randomly generated blocks-world problems, which are simply randomly-generated initial situations for a given number of blocks. For each random problem, a domain-specific routine is first applied to generate the set of alternative movements for each block. Then WATPLAN is applied for selecting sub-solutions and resolving conflicts. To compare with an incremental planner, an additional run is made for each test problem using TWEAK to combine sub-plans and resolve conflicts. The results are shown in Figure 9.1, where each datum is computed as the average result of 10 randomly generated problems for a given number of blocks. This test again demonstrates that with WATPLAN the computational cost of the combination phase is much lower than TWEAK.

9.2 Optimization-Based Plan Selection

As with constraint-based multiple-goal plan-selection problems, it is often the case that more than one plan exists for each goal when the plans are merged. A computational difficulty occurs due to the need to choose one plan for each goal so that after merging these plans, the total cost of the merged plan is minimized.

9.2.1 Examples

We start out by describing some example problem domains where the multiple-goal plan-selection problem is important.

Example 1: Manufacturing Process Planning. Much work has been done in the area of process planning for manufacturing, where the aim is to develop a plan of action for producing a machined part. As an example, consider the problem of using machining operations to make holes in a metal block. Several different kinds of hole-creation operations are available (twist-drilling, spade-drilling, gun-drilling, etc.), as well as several different kinds of hole-improvement operations (reaming, boring, grinding, etc.). Each time one switches to a different kind of operation or to a hole of a different diameter, one must mount a different cutting tool on the machine tool. If the same machining operation is to be performed on two different holes of the same diameter, then these two operations can be merged by omitting the task of changing the cutting tool. This and several other similar manufacturing problems are of practical significance (see [27, 63]).

Example 2: Query Plan Optimization. Another important area of problems related to planning is that of multiple-query optimization in database systems.
Let $Q = \{Q_1, Q_2, \ldots, Q_n\}$ be a set of queries to be processed. Associated with each query Q_i is a set of alternate access plans $\{\Pi_{i1}, \Pi_{i2}, \ldots, \Pi_{ik_i}\}$. Each plan is a set of partially ordered tasks that produces the answer to Q. For example, one task might be to find all employees in some department whose ages are less than 30, and whose salaries are over $50,000. Each task has a cost, and the cost of a plan is the sum of the costs of its tasks. Two tasks can be merged if they are the same, or if the result of evaluating one task reduces the cost of evaluating the other. The multiple-query optimization problem [117] is to find a global access plan by selecting and merging the individual plans so that the cost of the global plan is minimized.

Example 3: Common Sense Planning. Generating multiple plans may sometimes lead to better results even if the least costly plan is generated first. Consider a planning situation described by Wilensky [145]:

> John lives one mile from a bakery and one mile from a dairy. The two stores are 1.5 miles apart. John has two goals: to buy bread and to buy milk.

This time, however, let us add the fact (based on [145]) that

> John lives 1.25 miles from a large grocery store that sells both bread and milk.

The best plans for the individual goals involve two separate trips: one to the store and one to the dairy. However, this would require making two 2-mile trips for a total of 4 miles. The approach described in the previous section would allow them to be merged so that John could go directly from one store to the other (for a total trip of $1 + 1.5 + 1 = 3.5$ miles). A better plan, however, is to use the second-best plan for each goal (going to the grocery store). Even though taken separately these would generate a worse plan (two 2.5-mile trips for a total of 5 miles), they permit more significant merging when combined together (a single trip of $1.25 + 1.25 = 2.5$ miles). Thus, if the planners for the individual trips delivered more than one solution for each goal, this better plan could be found.

In all of the above domains more than one plan may be generated for each goal. This necessitates choosing among the plans available for each goal in order to find an optimal global plan. This problem in general is NP-hard, and in this chapter, we present a heuristic search technique that works quite well in practice.

9.2.2 Complexity

Our problem can be stated as follows. Given a set $\mathcal{G}$ of goals $g_i, i = 1, 2, \ldots k$, and given a set $\mathcal{P}_i$ of alternate plans for each goal set g_i, select one plan for each goal such that the resulting merged plan has a minimal cost.

This problem is NP-hard, even without considering the conflicts among the plans. We can prove this claim by a reduction from the *hitting set problem* [55], which is known to be NP-complete. The hitting set problem can be stated as follows:

> Given a collection C of subsets of a finite set S, and a positive integer $K \leq |S|$, is there a subset $S' \subseteq S$ with $|S'| \leq K$ such that S' contains at least one element from each subset in C?

The transformation to a multiple-plan selection problem can be done as follows. Every subset of S is a goal, and each element is a plan step. Two plan steps are mergeable if their corresponding elements are identical.

Despite the fact that the problem is NP-hard to solve, polynomial-time solutions do exist for several special cases of the merged plan existence and optimal merged-plan problems. One special case is where the number of different operator types is less than 3, assuming that plan steps of the same type can be merged. In this case, if no conflicting constraints are allowed to exist, the problem can be solved in polynomial time. For example, this would occur in Example 1 of Section 9.2.1 if there were only two different kinds of machining tools to be considered.

9.2.3 A Heuristic Algorithm for Plan Selection

Since the multiple-goal plan-selection problem is NP-hard, we resort to a heuristic method. Our algorithm will be based on *branch-and-bound*, introduced in Chapter 4.

A Branch and Bound Algorithm. Suppose that for each goal g_i we are given a set of plans $\mathcal{P}_i$ containing one or more plans for g_i. Suppose also that we are given a number of operator types, such that plan steps of the same type can be merged. For a set of plans $S = \{\Pi_i, i = 1, 2, \ldots\}$, we denote by COMBINE($S$) the plan after all conflicts are resolved in the set, using constraint-based techniques developed in Chapter 7. We denote by MERGE(S) the merged plan using either OPTIMALMERGE or APPROXMERGE algorithms of the last chapter.

Our state space is a tree in which each state at the i-th level is a plan for the goals $g_1, g_2, \ldots, g_i$. This tree is defined as follows (an example is shown in Figure 9.2):

1. the initial state is the empty set;
2. for each state S at level i of the tree, the children of S consist of all plans of the form MERGE(COMBINE($\{S, \Pi\}$)) such that $\Pi \in \mathcal{P}_{i+1}$ is a plan for g_{i+1}

Every state at level k is a final state, for these states are plans for the conjoined goal $G = \{g_1, g_2, \ldots, g_k\}$. A goal state is a final state F for which MERGE(COMBINE(F)) is minimal among all final states.

To search the state space, we use the best-first branch-and-bound algorithm shown below (for an overview of branch and bound, see Chapter 4). This algorithm maintains an active (or open) list A that contains all states

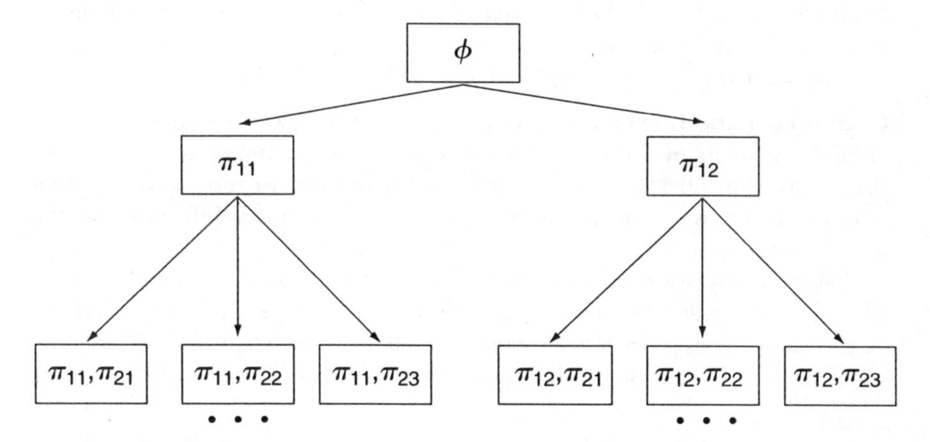

Fig. 9.2. An example state space. Here Π_{ij} is the j-th alternate plan for goal-set G_i

Table 9.2. The OPTIMALSELECT algorithm

Algorithm OPTIMALSELECT(Π)
Input: A set of plan sets $\mathcal{P}_i, i = 1, 2, \ldots, k$,
External Functions:
Children(S) returns the successor states of S;
$L(S)$ is a lower bound function, and $U(S)$ an upper bound function.
Output: A merged global plan.

```
1.    A := {∅};
      /* (A is the branch-and-bound active set)*/
2.    loop
3.      remove from A the state S for which L(S) is smallest;
4.      if S is a goal state then return(S);
5.      else
6.        C := Children(S);
7.        for each state S1 in A do
8.          for each state S2 in C do
9.          if U(S1) > L(S2) then
10.           remove S1 from A;
11.         else if U(S2) > L(S1) then
12.           remove S2 from C;
13.         end if
14.       A := A ∪ C
15.     end if
16.   repeat
```

---

eligible for expansion. To choose which member of A to expand next, the algorithm makes use of a function $L(S)$ that returns a *lower bound* on the costs of all descendants of S that are goal states. It always chooses for expansion the state $S \in A$ for which $L(S)$ is smallest.[1]

The algorithm OPTIMALSELECT is shown in Table 9.2.

Computing the Lower Bound. If $L(S)$ is a true lower bound on the costs of all descendants of S that are goal states, then L is admissible, in the sense that Algorithm OPTIMALSELECT will be guaranteed to return the optimal solution. Below we develop a lower bound function that is informed enough to reduce the search space dramatically in many cases.

Let S be any state at level i in the state space, and T be any child of S. Then T is the result of merging S with some plan $\Pi \in \mathcal{P}_{i+1}$. Let $N(\Pi, S)$ be the set of all steps in Π that *cannot be merged with steps* in S. This is the set of steps in Π that are of different operator types from any of those in S.

[1] The relationship between best-first branch and bound and the A* algorithm is well known [99]. The quantities $L(S)$, Cost(S), and $L(S)$ – Cost(S) used above are analogous to the quantities $f(S)$, G(S), and $h(S)$ used in A*.

We have

$$\text{Cost}(T) \geq \min_{\Pi \in \mathcal{P}_{i+1}} \text{Cost}(N(\Pi, S)) + \text{Cost}(S).$$

Similarly, if U is any child of T at level $i+2$, then

$$\text{Cost}(U) \geq \text{Cost}(S) + \max \begin{cases} \min_{\Pi \in \mathcal{P}_{i+1}} \text{Cost}(N(\Pi, S)), \\ \min_{\Pi \in \mathcal{P}_{i+2}} \text{Cost}(N(\Pi, S)). \end{cases}$$

By applying this argument repeatedly, we get

$$\text{Cost}(V) \geq \text{Cost}(S) + \max_{j>i} \min_{\Pi \in \mathcal{P}_j} \text{Cost}(N(\Pi, S)) \tag{9.1}$$

for every goal state V below S. Thus $L_1(S) = \text{Cost}(S)$ is an admissible lower bound function (this would correspond to using $h \equiv 0$ in the A* search algorithm). However, a better lower bound can be found from formula (9.1):

$$L_2(S) = \text{Cost}(S) + \max_{j>i} \min_{\Pi \in \mathcal{P}_j} \text{Cost}(N(\Pi, S))$$

L_2 is an admissible lower bound function that is better than L_1.

Computing Cost(S). The lower bound computation depends on the cost of a state resulting from merging a set of plans. There are several ways to estimate the cost.

1. Apply the dynamic programming algorithm OptimalMerge to the set of plans in state S, thereby obtaining an optimal cost; or
2. Apply the approximation algorithm ApproxMerge to the set of plans in state S, thereby obtaining a non-optimal cost;

If we follow method 1 above, then obviously we could obtain an accurate measure on the cost of a merged plan in each state, and the resultant lower bound heuristics L_1 and L_2 are accurate as well.

Improving the Lower Bound. By making more intelligent use of the information contained in the sets $N(\Pi, S)$, we can compute an even better lower bound function. The function $L_3(S)$ is described below.

Let S be a state at level i in the state space, and let j, k be two indexes that are higher than i: $j,\ k > i$. We say that $\mathcal{P}_j$ and $\mathcal{P}_k$ are S-*connected* if either of the following conditions holds:

- there are plans $\Pi \in \mathcal{P}_j$ and $Q \in \mathcal{P}_k$ such that $N(\Pi, S)$ and $N(Q, S)$ contain some steps that are mergeable, or
- there is a set $\mathcal{P}_h$ that is S-connected to both Π_j and Π_k.

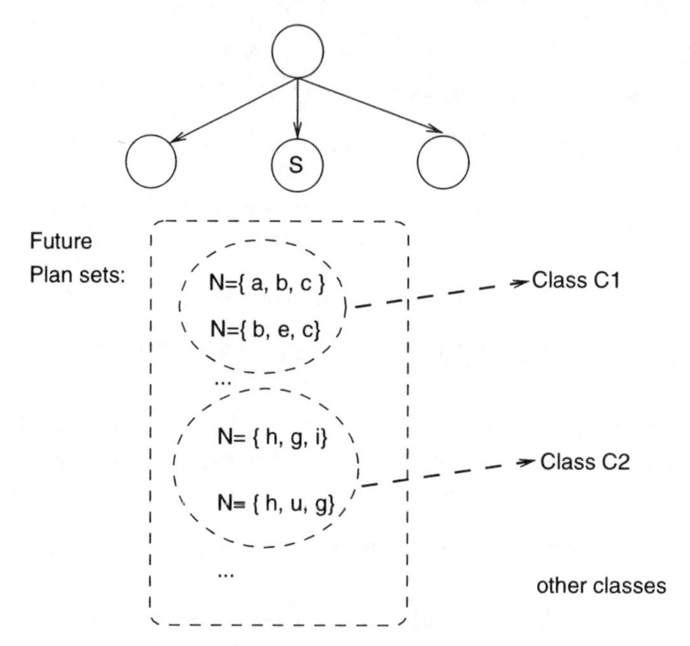

Fig. 9.3. Illustrating S-connected classes

S-connectedness is an equivalence relation, so we let $C_1(S), C_2(S), \ldots,$ be the equivalence classes. Therefore each class $C_k(S)$ contains one or more of the $\mathcal{P}_j$'s. An example is shown in Figure 9.3. We refer to these equivalence classes as *S-connectedness classes.*

Having done this, we can now define

$$L_3(S) = \text{Cost}(S) + \sum_j \max_{\mathcal{P}_k \in C_j(S)} \min_{\Pi \in \mathcal{P}_k} \text{Cost}(N(\Pi, S)) \tag{9.2}$$

It can easily be shown that L_3 is a lower bound on the cost of any descendant of S that is a goal state.

Computing an Upper Bound. The last factor in the heuristics of Algorithm OPTIMALSELECT is the upper bound $U(S)$ for each state S. Computation of the upperbound is ultimately related to whether one can quickly select one plan for each goal so that they can be consistently combined. This is a complex problem, as we have seen in Chapter 7. Without more domain knowledge, we cannot say much about whether we could succeed in finding a consistent set of plans to combine. However, in practical applications of the algorithm, as we see below in the manufacturing domain, strong knowledge is often available for guiding the selection process.

For now, we will make the simplifying assumption that *any* selection of plans can be, in one way or another, combined into one in which all conflicts are resolved. This assumption allows us to concentrate on the optimization aspect of the problem.

Let S be a state at level i, and let $\{\mathcal{P}_j, j = i+1, i+2, \ldots, k\}$ be the remaining plan sets. An upper bound estimate $U(S)$ can be constructed by taking a minimal cost plan Π_j from each set $\mathcal{P}_j$, and then computing the cost of a merged plan, using either OPTIMALMERGE or APPROXMERGE, of the plan set $\{S, \Pi_j, j = i+1, i+2, \ldots, k\}$. This value is a true upper bound on the optimal solution under state S because it corresponds to the cost of a leaf node in the subtree under S.

9.2.4 Examples

We now demonstrate the application of Algorithm OPTIMALSELECT in two domains.

A Process Planning Example. Consider an example from a machining domain (see Example 2 of Section 9.2.1). Suppose that the goal is to drill two holes h_1 and h_2 on a piece of metal stock. To make hole h_1, the plan is

Π_1 : `spade-drill`$(h_1) \mapsto$ `bore`(h_1).

To make hole h_2, however, there are two alternative plans:

$\Pi_{21:}$ `twist-drill`$(h_2) \mapsto$ `bore`(h_2).
$\Pi_{22:}$ `spade-drill`$(h_2) \mapsto$ `bore`(h_2).

Each time one switches to a different kind of machining operation, a different cutting tool must be mounted. Suppose that the relative costs are as follow: 1 for each tool change, 1 for each twist drilling operation, 1 for each boring operation, and 1.5 for each spade drilling operation. Then the costs of the individual plans are $\text{Cost}(\Pi_1) = \text{Cost}(\Pi_{22}) = 4.5$ and $\text{Cost}(\Pi_{21}) = 4$.

At level 0, the initial state $S_0 = \emptyset$, and the plan sets $\mathcal{P}_1 = \{\Pi_1\}$ and $\mathcal{P}_2 = \{\Pi_{21}, \Pi_{22}\}$ are S-connected since they share tool-changing operations for boring and spade-drilling.

At level 1, the state-space has only one state, namely $S_1 = \{\Pi_1\}$. $N(\Pi_{21}, S_1) = \{\texttt{twist-drill}(h_2)\}$ and $N(\Pi_{22}, S_1) = \emptyset$. Thus,

$$L_3(S_1) = \text{Cost}(S_1) + \min\{\text{Cost}(N(\Pi_{21}, S_1)), \text{Cost}(N(\Pi_{22}, S_1))\} = 4.$$

At level 2, the two successor states of S_1 are:

$$\begin{aligned} S_{21} &= \{\text{MERGE}(\text{COMBINE}(\{\Pi_1, \Pi_{21}\}))\}, \\ S_{22} &= \{\text{MERGE}(\text{COMBINE}(\{\Pi_1, \Pi_{22}\}))\} \end{aligned}$$

Their heuristic estimates are $L_3(S_{21}) = 7$, $L_3(S_{22}) = 6.5$. Thus, the optimally merged plan is

$S_{22} =$ `spade-drill`$(h_1, h_2) \mapsto$ `bore`$(h_1$ and $h_2)$.

A Household Domain. We now demonstrate Algorithm OPTIMALSELECT on the "bread and milk" example, Example 3, from Section 9.2.1. A graphical illustration is shown in Figure 9.4. The plans for the goal have(Bread) are

Π_{11}: go(Home, Bakery) $\mapsto$ buy(Bread), $\mapsto$ go(Bakery, Home);
Π_{12}: go(Home, Grocery) $\mapsto$ buy(Bread), $\mapsto$ go(Grocery, Home).

The plans for the goal have(Milk) are

Π_{21}: go(Home, Dairy) $\mapsto$ buy(Milk) $\mapsto$ go(Dairy, Home).
Π_{22}: go(Home, Grocery) $\mapsto$ buy(Milk) $\mapsto$ go(Grocery, Home).

We now trace the operation of Algorithm OPTIMALSELECT.
At level 1, there are two states,

$$S_1 = \{\Pi_{11}\}, S_2 = \{\Pi_{12}\}.$$

Taking the distance between any two locations as the cost of going from one to the other, we have

$$\text{Cost}(S_1) = 2, \text{Cost}(S_2) = 2.5.$$

The heuristic function values are

$$L_3(S_1) = 2 + \min\{\text{Cost}(\{\texttt{buy}(\text{Milk})\}), \text{Cost}(\Pi_{22})\} = 2, \text{ and}$$

$$L_3(S_2) = 2.5 + \min\{\text{Cost}(\Pi_{21}), \text{Cost}(\{\texttt{buy}(\text{Milk})\})\} = 2.5.$$

Thus, the algorithm will expand S_1 next.
There are two successors of S_1.

$$T_1 = \{\Pi_{11}, \Pi_{21}\}; \qquad T_2 = \{\Pi_{12}, \Pi_{22}\}.$$

The merged plan T_1 corresponds to going to the dairy to buy milk, going from the dairy to the bakery to buy bread, and finally going home from the

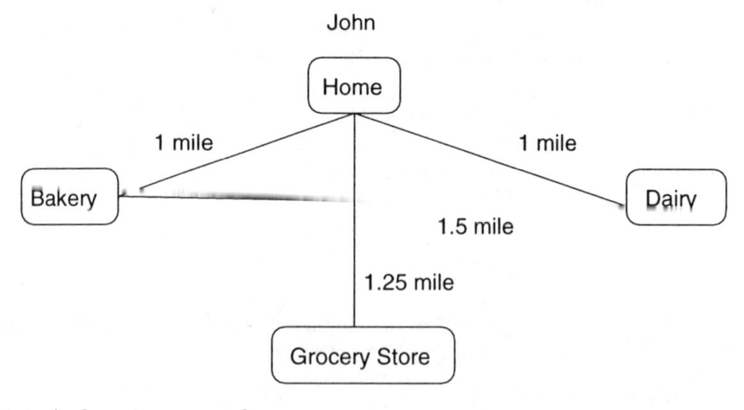

Fig. 9.4. A shopping example

bakery. The cost of this plan is $\text{Cost}(T_1) = 1 + 1.5 + 1 = 3.5$. T_2 corresponds to going to the bakery and the grocery store on separate trips, giving rise to a cost of 4.5. Since both are more costly than S_2, S_2 will be expanded next. One successor of S_2 combines and merges Π_{12} with Π_{22}, yielding the plan

$$\texttt{go}(\text{Home, Grocery}), \mapsto \texttt{buy}(\text{Milk and Bread}), \mapsto \texttt{go}(\text{Grocery, Home}).$$

In this case, this plan is the optimal merged plan, with a cost of 2.5.

9.2.5 Empirical Results in a Manufacturing Planning Domain

To test the effectiveness of the heuristics, experiments are done by Nau and Yang with the algorithm using the EFHA process planning system [136], a domain-dependent planner based on the earlier SIPS process planner, developed by Nau et al. at University of Maryland [98]. The parameters involved in the generation of alternate plans are varied to make sure they would not be overly uniform. Particular attention are paid to ensure that the test problems are typical of those found in manufacturing applications.

The target problem is to find a least-cost plan for making several holes in a piece of metal stock. Specifications for 100 machined holes are generated, randomly varying various hole characteristics such as depth, diameter, surface finish, locational tolerance, etc. EFHA is executed to produce at most 3 plans for each hole, and it found plans for 81 of the holes (for the other 19 the machining requirements were so stringent that EFHA could not produce any plans using the machining techniques in its knowledge base).

The distributions of the hole characteristics are chosen so that the plans generated for the holes would have the following characteristics:

1. a wide selection of plans, rather than lots of duplicate plans for different holes;
2. not very many holes having an "obviously best" plan (i.e., a plan that is a sub-plan of all the other plans for that hole);
3. lots of opportunities to merge steps in different plans;
4. a large number of "mergeability tradeoffs" in choosing which plan to use for a goal. That is, it is nontrivial to decide which alternative plans to select in order to minimize their costs.

The results of the experiments are shown in Table 9.3. Each entry in the table represents an average result over 450 trials. Each trial was generated by randomly choosing m of the 81 holes (duplicate choices were allowed), invoking Algorithm OPTIMALSELECT on the plans for these holes using the lower bounding function L_3, and recording how many nodes it expanded in the search space. The total cost of each plan was taken to be the sum of the costs of the machining operations in the plan and the costs for changing tools.

The performance of the algorithm are quite good—especially since the test problem was chosen to be significantly more difficult than the kind of problem that would arise in real-world process planning. In real designs,

Table 9.3. Experimental results for Algorithm 4 using L_3

Number of holes m	Nodes in the search space	Nodes expanded
1	2	1
2	10	2
3	34	3
4	98	4
5	284	6
6	852	9
7	2372	12
8	6620	16
9	19480	22
10	54679	28
11	153467	38
12	437460	51
13	1268443	61
14	3555297	86
15	9655279	110
16	29600354	170
17	80748443	223
18	250592571	250

designers would normally specify holes in a much more regular manner than our random choice of holes, making the merging task easier. For example, when merging real-world process plans, many of the mergeability tradeoffs mentioned earlier would not actually occur; and without such tradeoffs, the complexity of the algorithm is polynomial rather than exponential.

9.3 Open Problems

Combining OPTIMALSELECT **with** WATPLAN**.** One problem which we did not address in this chapter is how to extend the merging-based plan-selection technique to include constraint-based plan selection in WATPLAN. WATPLAN can be seen as a satisficing method, for selecting a plan for each goal set such that a consistent combination can be obtained. OPTIMALSELECT, on the other hand, is an optimization algorithm for selecting plans based on quality. To combine the two algorithms, one could use plan merging as a basis for designing a value-selection heuristic in WATPLAN. This heuristic prefers a low-cost plan whenever choices are available. The resultant solution found in the end may not be the one with absolutely the lowest cost, but it will be a good one.

Interleaving Plan Selection with Plan Generation. The problem model we have so far relied on assumes that a set of alternative plans is given for each goal-set. It is possible that to generate the alternatives, one has to spend vast computational resources. Therefore, it would be more efficient if one could select plans as they are generated, rather than waiting till all plans are constructed. We suspect that to do this one needs to have some special kind of knowledge about the problem domain, so that even before a plan is completely generated, some prediction could be made regarding whether the plan can be fruitfully combined and merged with others.

9.4 Summary

The multiple-goal plan-selection problem is NP-hard. Constraint-based as well as a branch-and-bound heuristic-search algorithms are developed to find conjoined plans. An admissible heuristic, and several variants, are proposed for the merging-based multiple-goal plan-selection problem to show that this search can find optimal plans. Empirical results are shown demonstrating that in an interesting class of automated manufacturing problems, the heuristic algorithm performs quite well, still growing exponentially but by a very small factor.

9.5 Background

Discussion on consistency-based multiple-goal plan selection problem was partly based on an Artificial Intelligence journal article by Yang [154]. Discussion on merging-based multiple-goal plan selection problem is adapted from Yang's PhD dissertation [150], part of which was reported in a Computational Intelligence journal article by Yang, Nau, and Hendler [158]. Ephrati and Rosenschein presented an integrated framework for distributed planning, which could handle both conflicts and operator-overlapping interactions while a global plan is being generated [42]. A complete description of the application of merging-based multiple-goal plan selection methods to process planning is described in [76].

Part III

Hierarchical Abstraction

Overview

A superior strategist develops a plan
Before committing the entire force.

An inferior strategist commits the entire force
Before developing a plan

(Sun Tzu, *The Art of Strategy*, Chapter Four)

In this last part of the book we focus on hierarchical planning with abstraction. Planning with abstraction refers to the methodology in which an abstract version of a plan is first developed by concentrating on the most important aspects of a problem. After committing to the abstract plan, a refinement process will then follow to complete all the remaining details.

Planning with abstraction has had a long tradition in Artificial Intelligence, but the development of this subject into a vigorous discipline has been rather recent. In Chapter 10 we give a general introduction to abstraction, providing a basic computational framework in which we could design algorithms and perform analysis. In this chapter we also describe an array of properties with which we could judge objectively the quality of an abstraction hierarchy. With the formalism at hand, we then turn around and develop an algorithm for automatically generating abstraction hierarchies (Chapter 11). This work is built on previous achievements in the field, which we review in detail in this chapter. In Chapter 12 we examine another type of hierarchical planner, using hierarchical task networks. We show that some important formal properties are lost with this type of planner, and we also illustrate how to remedy these losses. Finally, in Chapter 13, we present a method for automatically abstracting the effects of planning operators. The method is based on a learning algorithm, which is able to find an appropriate balance between the completeness of a planner and its computational efficiency.

10. Hierarchical Planning

Ever since the conception of Artificial Intelligence, *hierarchical problem solving* has been used as a method to reduce the computational cost of planning. The idea of hierarchical problem-solving, a well-accepted one, is to distinguish between goals and actions of different degrees of importance, and solve the most important problems first. Its main advantage derives from the fact that by emphasizing certain activities while temporarily ignoring others, it is possible to obtain a much smaller search space in which to find a plan.

As an example, suppose that in the household domain we would like to paint the ceiling white. Initially the number of conditions to consider may be overwhelming, ranging from the availability of various supplies, the suppliers for equipments and tools, to the position of the agent, the ladder, and the state of the ceiling. However, we could obtain a more manageable search space by first concentrating on whether we have the paint, the ladder, and a brush. Once a plan is found we then consider how to *refine* this plan by considering how to get to the rooms where each item is located. The process repeats until a full-blown plan is finally found.

Today, a large number of problem-solving systems have been implemented and studied based on the concept of hierarchical planning. They include GPS [101], ABSTRIPS [113], LAWLY [121], NOAH [114], NONLIN [132], MOLGEN [128], SOAR [138], SIPE [146], and ABTWEAK [160].

In this part of the book, we provide a general introduction of hierarchical planning, exploring the uses of hierarchies in planning in several contexts. We first provide a generic introduction to hierarchical planning, in which a plan is gradually refined by considering increasing levels of detail. We then specifically consider two ways in which details are inserted in a plan. The first, *precondition-elimination abstraction*, mimics the human intuition of exploring and solving subgoals in order of importance, with the most important goals solved first.

A second method of hierarchical planning that we consider is called *hierarchical task-network planning* (HTN), where a planning problem and operators are organized into a set of tasks. A high-level task can be *reduced* to a set of ordered lower-level tasks, and a task can be reduced in several ways. The reduction functions as a mapping from a task to a set of subtasks, defining the knowledge about how one can obtain a detailed plan from an abstract one. These two methods are complementary—each can be very effective in solving a different type of problem.

10.1 A Hierarchical Planner

10.1.1 Algorithm

The intuition behind the operation of a hierarchical planner is shown in Figure 10.1. In this figure there are three levels of abstraction, an abstract level, an intermediate level and a concrete level. Each dashed box represents a problem-solver at a given level. A planning problem is first abstracted and solved at the most abstract level. The solution obtained at this level, an *abstract plan*, is taken as the input to a problem-solver at the next level. The process ends when a concrete-level solution is found.

In general, the abstraction levels could range from a single level to multiple levels. The former is identical to problem-solving without any abstraction. A top-level description of a hierarchical partial-order planner is shown in Table 10.1.

For a given problem, Algorithm HIPLAN starts with an initial plan Π_0 representing the start-step/finish-step pair. It then iterates through a cycle

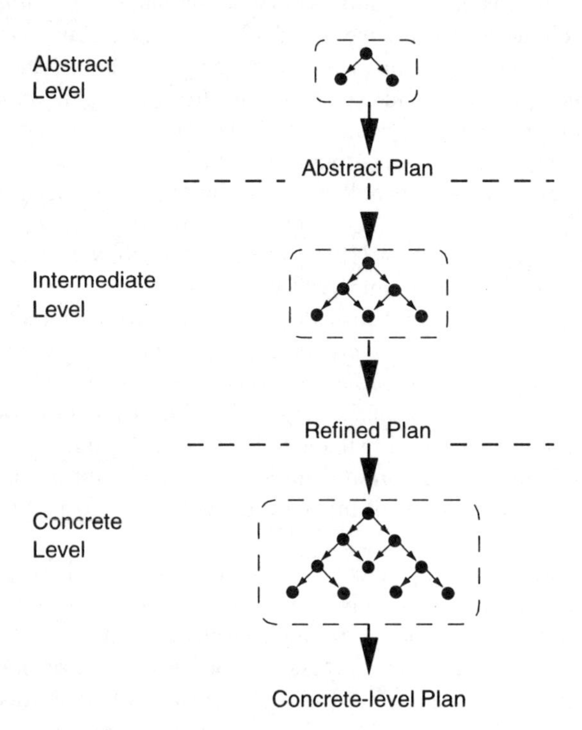

Fig. 10.1. Illustrating hierarchical planning

Table 10.1. The HiPlan algorithm

Algorithm HiPlan(Π_0)
Input: An initial abstract plan Π_0.
External Function: Refine(Π, i) returns a set of refinements of plan at abstraction level i.
Output: A correct plan consisting of concrete-level plan steps; or Fail.

```
1. OpenList := {Π0};
2. repeat
3.   Π := Remove(OpenList);
4.   if Level(Π) = 0 then
5.     if Correct(Π) then
6.       return(Π);
7.     end if
8.   else
9.     Level(Π) := Level(Π) − 1;
10.    Successors := Refine(Π, Level(Π));
11.    OpenList := Insert(Successors, OpenList);
12.  end if
13until OpenList = ∅;
14return(Fail);
```

of *refinement*, where in each iteration, a plan is chosen from the OpenList (Step 3), and if it is not a correct concrete-level plan, it will be lowered to the next level down (Step 9). The plan at this new level is likely to be temporarily incorrect, since some plan steps may be left out. To account for these missing components a *refinement* process (Step 10) will be initiated, in which more plan steps are filled in. The algorithm stops with a correct plan when it finds a plan Π consisting solely of concrete-level plan steps, and is correct (Steps 5,6).

In the algorithm, the function Refine(Π, i) is intentionally left unspecified. Depending on the type of hierarchical planning used, we could specify this function either *explicitly* or *implicitly*. In the next section, we conduct a case study where this function is specified *implicitly*.

10.1.2 Precondition-Elimination Abstraction

Precondition-elimination abstraction systems depend on a formulation of a problem domain as a multi-layered hierarchy. The form of abstraction hierarchy largely determines the effectiveness of problem-solving, and not every hierarchy will lead to an improvement in efficiency.

In Abstrips [113], Sacerdoti developed an elegant means for specifying abstract problem descriptions by assigning *criticality values*. Here we follow

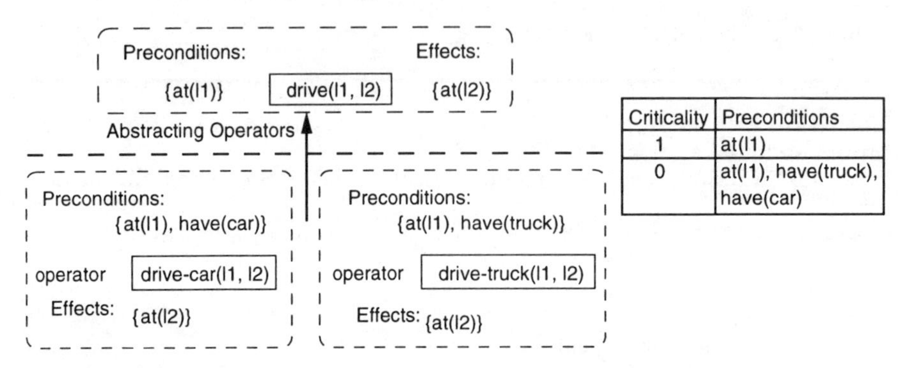

Fig. 10.2. Two operators with the same abstract representation

the same approach, but do not restrict our planning algorithm to a total-order planner such as ABSTRIPS.

Defining a Hierarchy. Consider assigning the integers between 0 and $n-1$, for some finite value n, to preconditions. A problem description at level-i is obtained by eliminating all literals having criticality less than i from the precondition of an operator. This procedure defines an n-level planning system — a triple $\Sigma = (L, \mathcal{O}, crit)$, where L is the set of literals used in describing the problem domain, and $\mathcal{O}$ is the set of operators. *crit* is a function mapping preconditions to non-negative integers:

$$crit : \bigcup_{\alpha \in \mathcal{O}} \text{Pre}(\alpha) \rightarrow \{0, 1, \ldots, n-1\}.$$

Level 0 is called the *concrete level*, and level $n-1$ the most abstract level. The level of a plan is denoted as $\mathsf{Level}(\Pi)$.

Intuitively, *crit* is an assignment of criticality values to each literal appearing in the precondition of an operator. Let α be an operator. We take $\mathsf{Abs}(i, \text{Pre}(\alpha))$ to be the set of preconditions of α which have criticality values of at least i:

$$\mathsf{Abs}(i, \text{Pre}(\alpha)) = \{p \mid p \in \text{Pre}(\alpha) \textit{ and } crit(p) \geq i\}.$$

$\mathsf{Abs}(i, \alpha)$ is operator α with preconditions $\mathsf{Abs}(i, \text{Pre}(\alpha))$ and effects $\text{Eff}(\alpha)$.

Any abstract operator α may be the result of abstracting more than one lower-level operator. Figure 10.2 shows an example. Let $\alpha_1, \alpha_2, \ldots, \alpha_k$ all have the same abstract operator representation at level i, then the *cost* of the abstract operator is the *minimum* of the costs of all operators $\alpha_i, i = 1, 2 \ldots k$. For every abstract operator α at level i, we use $\text{NextLevel}(i, \alpha)$ to denote the set of $i-1$ level operators, such that every operator in the set can be abstracted to α at level i.

Finally, let the set of all such $\mathsf{Abs}(i, \alpha)$ be $\mathsf{Abs}(i, \mathcal{O})$. This defines a *planning problem description* on each level i of abstraction. The domain-description language and the planning operators are specified as:

$$\langle L, \ \mathsf{Abs}(i, \mathcal{O}) \rangle$$

The same definition applies to plan steps; for a step **s** in a plan at the concrete level, level 0, the preconditions of an abstract step at the i-th level are given by

$$\mathsf{Abs}(i, \text{Pre}(\mathbf{s})) = \{p \mid p \in \text{Pre}(\mathbf{s}) \ \textit{and} \ \textit{crit}(p) \geq i\}$$

The effects of $\mathsf{Abs}(i, \mathbf{s})$ remain the same, and the cost of $\mathsf{Abs}(i, \mathbf{s})$ is defined the same as the cost of an abstract operator.

In our definition of an abstraction hierarchy we have restricted our attention only to the preconditions. This type of abstraction has been called *relaxed models*, since by ignoring certain preconditions, a planning problem is simplified, and the number of solutions to a planning problem increases. If one further abstracts the effects of operators then in fact one can obtain a much smaller search space for a forward-chaining planner, and the resultant model is called *reduced model.* Knoblock [80] presents an excellent discussion of the two models, his conclusion being that most formal results obtained in one model are applicable to the other.

A Tower of Hanoi Example. We illustrate the specification of an abstraction hierarchy in the 3-disk version of the Tower of Hanoi (TOH) domain. As shown in Figure 10.3, initially all three disks are on Peg1. Disks can be moved one at a time, under the restriction that no disk of a larger size can rest on a smaller disk. The goal is to place all three disks on Peg3.

We can represent the domain with four predicates, Ispeg, Onsmall, Onmed, and Onlarge. The concrete-level operators in this domain are shown in Table 10.2. If a hierarchy is built by assigning a distinct criticality value to each of the predicates, then 24 different hierarchies exist. We show one criticality assignment in Table 10.3. This assignment has often been used to demonstrate the advantages of abstraction, most notably by Knoblock [80]. For ease of exposition, we use ILMS to represent the hierarchy.

The initial state of this problem is

$$\begin{gathered}\text{Ispeg(Peg1)}, \text{Ispeg(Peg2)}, \text{Ispeg(Peg3)},\\ \text{Onlarge(Peg1)}, \neg\text{Onlarge(Peg2)}, \neg\text{Onlarge(Peg3)},\\ \text{Onmed(Peg1)}, \neg\text{Onmed(Peg2)}, \neg\text{Onmed(Peg3)},\\ \text{Onsmall(Peg1)}, \neg\text{Onsmall(Peg2)}, \neg\text{Onsmall(Peg3)}\end{gathered}$$

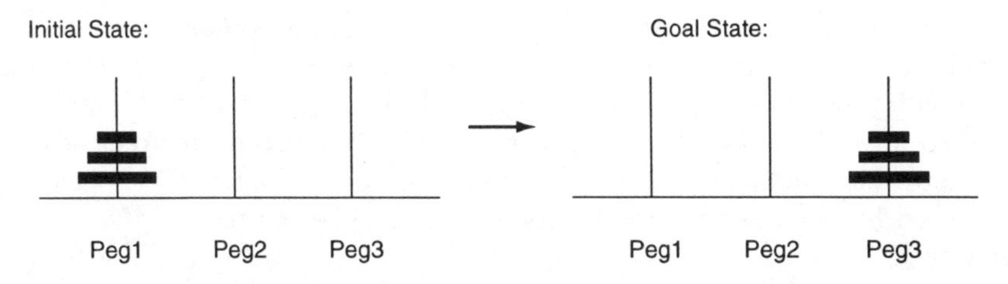

Fig. 10.3. A Towers of Hanoi domain

Table 10.2. Operator definitions in the TOH domain

Preconditions		Effects	Cost
`movelarge`($?x$, $?y$)			
Ispeg($?x$), Onlarge(?x), ¬Onmed($?x$), ¬Onmed($?y$), ¬Onsmall($?x$), ¬Onsmall($?y$)	Ispeg($?y$),	Onlarge($?y$), ¬Onlarge($?x$)	1.0
`movemed`($?x$, $?y$)			
Ispeg($?x$), Onmed(?x), ¬Onsmall($?x$), ¬Onsmall($?y$)	Ispeg($?y$),	Onmed($?y$), ¬Onmed($?x$)	1.0
`movesmall`($?x$, $?y$)			
Ispeg($?x$), Onsmall(?x)	Ispeg($?y$),	Onsmall($?y$), ¬Onsmall($?x$)	1.0

Table 10.3. A criticality assignment in the TOH domain

Criticality	Predicates
2	Ispeg, Onlarge
1	Onmed
0	Onsmall

And a goal is

$$\text{Onlarge(Peg3)}, \text{Onmed(Peg3)}, \text{Onsmall(Peg3)}$$

For the ILMS hierarchy, at the most abstract level, the goal is

$$\text{Onlarge(Peg3)}$$

And the level-2 abstract operators are shown in Table 10.4.

The TOH domain is of interest in the study of abstraction because of a number of problem features. A planning problem in this domain typically consists of conjunctive goals, and the solution length increases exponentially with the number of goals. Many non-hierarchical planners consider this problem hard to solve. In addition, there exists a clear distinction of different degrees of importance among the domain conditions, and the interactions between these conditions can be carefully controlled to generate meaningful results, as we will see later.

Table 10.4. Abstract operator definitions in TOH domain

Preconditions	Effects	Cost
Abs(**2**, movelarge(?x, ?y))		
Ispeg(?x), Ispeg(?y), Onlarge(?x)	Onlarge(?y), ¬Onlarge(?x)	1.0
Abs(**2**, movemed(?x, ?y))		
Ispeg(?x), Ispeg(?y)	Onmed(?y), ¬Onmed(?x)	1.0
Abs(**2**, movesmall(?x, ?y))		
Ispeg(?x), Ispeg(?y)	Onsmall(?y), ¬Onsmall(?x)	1.0

A solution at the most abstract level – level-2 of the ILMS hierarchy – is

$$\Pi_1 = \mathbf{s}_{init} \mapsto \texttt{movelarge}(\text{Peg1}, \text{Peg3}) \mapsto \mathbf{s}_{finish}$$

At this level there are also other more costly abstract solutions. One of them is

$$\Pi_2 = \mathbf{s}_{init} \mapsto \texttt{movelarge}(\text{Peg1}, \text{Peg2}) \mapsto \texttt{movelarge}(\text{Peg2}, \text{Peg3}) \mapsto \mathbf{s}_{finish}$$

An optimal solution with seven steps can be found by refining Π_1:

$$\begin{array}{c}
\mathbf{s}_{init} \\
\downarrow \\
\texttt{moveS}(\text{Peg1}, \text{Peg3}) \\
\downarrow \\
\texttt{moveM}(\text{Peg1}, \text{Peg2}) \\
\downarrow \\
\texttt{moveS}(\text{Peg3}, \text{Peg2}) \\
\downarrow \\
\texttt{moveL}(\text{Peg1}, \text{Peg3}) \\
\downarrow \\
\texttt{moveS}(\text{Peg2}, \text{Peg1}) \\
\downarrow \\
\texttt{moveM}(\text{Peg2}, \text{Peg3}) \\
\downarrow \\
\texttt{moveS}(\text{Peg1}, \text{Peg3}) \\
\downarrow \\
\mathbf{s}_{finish}
\end{array}$$

A Simple Travel Domain. We next consider a simple travel domain, where the task is to determine a route for driving from one city to the next (see Figure 10.4). To go to a destination city, it may be necessary to stop over in a number of intermediate cities, and within each of these cities, it may be

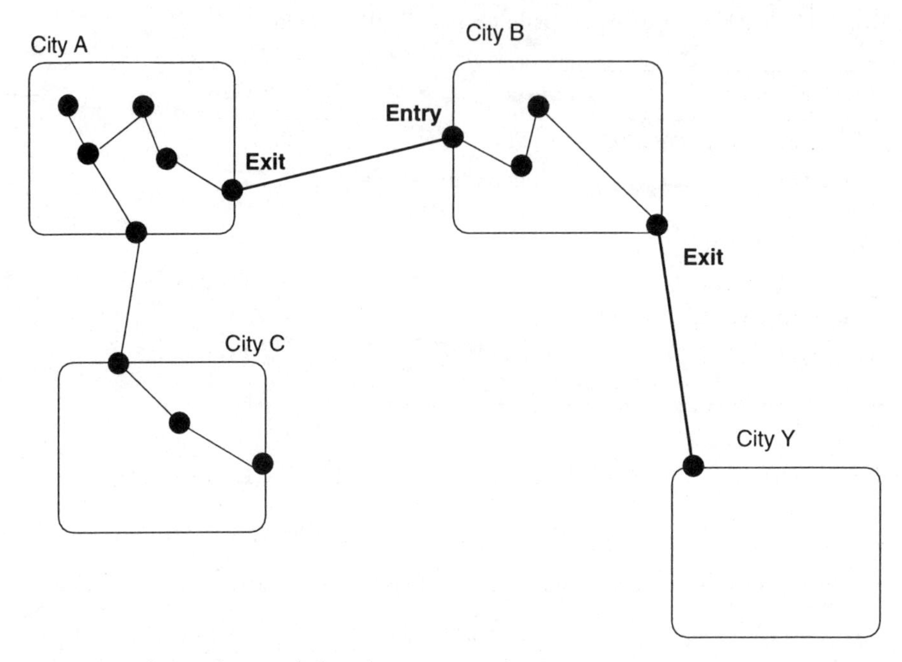

Fig. 10.4. A simple travel domain

necessary to find a route to connect the entry point to an exit point. The operators of this domain are listed in Table 10.5.

An initial state of a planning problem can be specified as

$$\text{In-City}(A), \text{At-Loc}(\text{Exit}_A), \text{Connected}(\text{Exit}_A, A, \text{Entry}_B, B),$$
$$\text{Connected}(\text{Exit}_B, B, \text{Entry}_C, C), \ldots, \text{Connected}(\text{Exit}_X, X, \text{Entry}_Y, Y).$$

A goal might be

$$\text{In-City}(Y)$$

To solve this problem, an abstract plan corresponds to an inter-city route, and a concrete-level plan corresponds to one that has all intra-city routes planned out.

Table 10.5. Operators in a simple travel domain

Preconditions	*Effects*	*Cost*
`drive-bet-cities`($?c_1$, $?c_2$)		
Connected($?exit, ?c_1, ?entry, ?c_2$), In-City($?c_1$), At-Loc($?exit$)	In-City($?c_2$), At-Loc($?entry$), ¬In-City($?c_1$), ¬At-Loc($?exit$)	10.0
`drive-in-city`($?c$, $?l_1$, $?l_2$)		
In-City($?c$), At-Loc($?l_1$)	At-Loc($?l_2$), ¬At-Loc($?l_1$)	1.0

10.2 Specifying Refinement

As with ABSTRIPS, HIPLAN performs planning level by level, in a top-down fashion. Once a plan is found at a given abstraction level i, all preconditions at level $i-1$ — the immediate lower-level — are put back into the preconditions of plan steps. As a result of the new preconditions, the plan may not be correct at level $i-1$. This calls for a process to *refine* the plan, to make it correct again at level $i-1$. Below, we discuss the refinement process in two settings, with forward-chaining, total-order planners and with backward-chaining, partial-order planners.

10.2.1 Forward-Chaining, Total-Order Refinement

A total-order plan Π at abstract level i can be considered as a skeletal plan for the next level down. Let an i-th-level plan Π be a sequence of plan steps

$$\mathbf{s}_0 = init \mapsto \mathbf{s}_1 \mapsto \mathbf{s}_2 \mapsto \ldots goal = \mathbf{s}_r$$

At the next lower level $i-1$, between every pair of steps $(\mathbf{s}_j,\ \mathbf{s}_{j+1})$ there appears a "gap" problem due to the introduction of new preconditions at level $i-1$. In the simple travel example introduced in the last section, an abstract-level plan consists of inter-city driving specifications only. When taken down to the intra-city level, however, between every "entry" point to a city and an "exit" point from the same city, a route has yet to be specified. The route is the gap problem. Figure 10.5 shows an example.

Each of the gap problems is called a *subproblem*. Before we can call a problem-solver for solving the problem, we must fully specify its initial and goal states. The goals for the subproblems are easy to specify; they simply consist of all preconditions of the next step. The initial states, however, might depend on the plans found for all gap subproblems *before* the current step.

To address this issue, we follow the following *length-first* method of refinement [113]. The subproblems can be solved in a left-to-right scanning operation. Starting from the left, we number the subproblems from 1 through r (see Figure 10.5). The initial state of the first subproblem is simply the initial

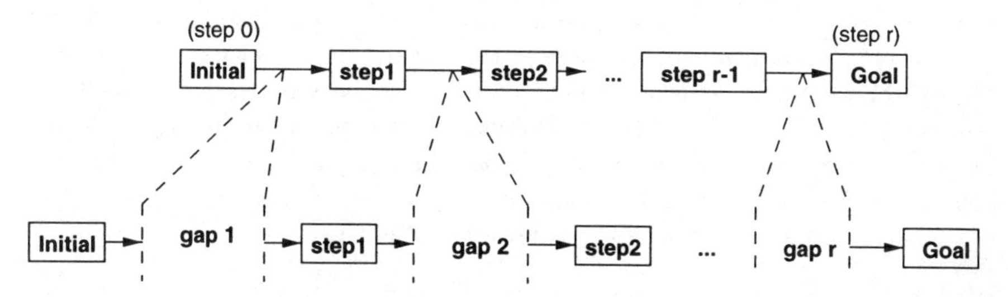

Fig. 10.5. Refinement of a total-order plan

state of the concrete problem. The goal is the set of all preconditions of the first plan step, $\mathbf{s}_1$, at level $i-1$:

$$goal_1 = \text{Pre}(\mathsf{Abs}(i-1,\ \mathbf{s}_1))$$

A solution to this problem is obtained by a call to a forward-chaining, total-order planner $\Pi_1 = \text{ToPlan}(init,\ goal_1)$. For a specification of ToPlan see Chapter 2.

To obtain the initial state for the next subproblem, we apply the plan $[\Pi_1 \mapsto \mathbf{s}_1]$ to the initial state, and obtain a new initial state $init_2$, for the next subproblem. In the j-th iteration, suppose that all previous $j-1$ subproblems have already been solved. Let the solutions be $\Pi_m, m = 1, 2, \ldots j-1$, respectively. Then for the j-th subproblem, the initial state is obtained by applying the plan

$$\Pi_1 \mapsto \mathbf{s}_1 \mapsto \Pi_2 \mapsto \mathbf{s}_2 \mapsto \Pi_3 \ldots \mathbf{s}_{j-1}$$

to the original initial state. This process continues until the final goal is achieved.

A formulation of the above length-first refinement algorithm TotalRefine() is shown in Table 10.6. In this algorithm, $\mathcal{O}_b$ is the set of concrete-level operators. ApplyPlan is a function that applies a plan to a state, and returns the resulting state.

10.2.2 Backward-Chaining, Partial-Order Refinement

For partial-order refinements, there is a higher degree of continuity between planning at one level and planning at the next. For any given plan Π at an abstraction level i, all plan steps are first replaced by their corresponding $(i-1)$-th-level steps. If an abstract step $\mathbf{s}$ has several corresponding $(i-1)$-th-level steps, as specified in set $\text{NextLevel}(i-1, \mathbf{s})$, one of the steps is chosen and the rest saved as backtrack points. Subsequently, the new preconditions will be planned for at level $i-1$. The algorithm PartialRefine is shown in Table 10.7.

The continuity between planning at different levels of abstraction is a direct consequence of the flexibility offered by partial-order plans. Because the order in which a subgoal is achieved is not committed to until necessary, given two subgoals, a partial-order planner can derive the same plan by working on either subgoal first. In abstraction, this means that the rigid, left-to-right, length-first refinement of a total-order plan can be relaxed dramatically by allowing any open-precondition to be worked on without sacrificing the ability to find a refinement. In this way, an abstract partial-order planner works in the same way as a single-level planner, with abstraction providing a more intelligent ordering in which to achieve its subgoals.

Table 10.6. The TOTALREFINE algorithm for precondition-elimination abstraction

---

Algorithm TOTALREFINE(Π)
Input: A plan Π at abstraction level i.
Output: A plan at abstraction level $i-1$, or Fail.

```
1.    i := crit(Π);
2.    for each step s ∈ Steps(Π) do
3.       replace s by a step in NextLevel(i − 1, s);
         /* if there are several steps in NextLevel(i − 1, s),
            pick one, and keep the rest as backtrack points. */
4.    end for
5.    let init be the initial state in Π;
6.    let O be Abs(i − 1, O_b);
         /* O_b are the concrete-level operators */
7.    let Π consist of r + 1 steps (initial state is step 0);
8..   for j = 1 to r do
9.       Π_j := TOPLAN(init, Pre(Abs(i − 1, s_j)));
         /* if there are several solutions, pick one,
            and keep the rest as backtrack points. */
10.      if Π_j = Fail then
11.         return(Fail);
12.      else
13.         init := APPLYPLAN(Π_j, init);
14.      end if
15.   end for
16.   REFINEMENT := s_init↦Π_1↦s_1 ... Π_r↦s_finish;
17.   return(REFINEMENT);
```

Table 10.7. The PARTIALREFINE algorithm for precondition-elimination abstraction

---

Algorithm PARTIALREFINE(Π)
Input: A plan Π at abstraction level i.
Output: A plan at abstraction level $i-1$, or Fail.

```
1.    i := crit(Π);
2.    for each step s ∈ Steps(Π) do
3.       replace s by a lower-level step in NextLevel(i − 1, α);
         /* if there are several steps in NextLevel(i − 1, s),
            keep the rest as backtrack points. */
4.    end for
5.    let O be Abs(i − 1, O_b);
6.    Solution := POPLAN(Π);
         /* for a description of the POPLAN algorithm see Chapter 2 */
7.    return(Solution);
```

10.3 Properties of an Abstraction Hierarchy

Given a problem domain represented by a description language L, a set of concrete-level operators $\mathcal{O}_b$ and initial and goal states, there are many possible abstraction hierarchies that one can construct. With precondition-elimination hierarchies, if L contains N literals, and if we place one literal on each level of the abstraction hierarchy, we could obtain a total of $N!$ abstraction hierarchies. Out of this many candidates, which one is good? This question has intrigued many planning researchers.

One way to answer the above question is to define a *property* of an abstraction hierarchy in terms of the relationship between an abstract plan and a concrete-level plan. The properties could presumably be linked to efficiency-gains in planning using the hierarchies. If a hierarchy satisfies a certain property, then we could classify it as a "good" one. This is the approach we will follow in this section.

10.3.1 Existence-Based Properties

Does the existence of an abstract solution guarantee the existence of a solution at any lower level? Or, is the converse true of an abstraction hierarchy? In this section, we first review two properties on abstraction hierarchies based on existence of plans.

Upward Solution Property. Consider a two-level hierarchy. Suppose that, whenever there exists any concrete-level plan Π_c, there exists an abstract plan Π_a. Then the hierarchy is said to satisfy the *upward solution property.*

In general, suppose that we are given an n-level abstraction hierarchy, with level $n-1$ being the most abstract and level 0 the most concrete level. Let i be an integer between 0 and $n-2$.

Definition 10.3.1 (Upward Solution Property). *Whenever any i-th-level solution Π_i exists, there exists an abstract solution Π_{i+1} at level $i+1$.*

By this definition, if a concrete-level solution Π_0 exists then there exists a sequence of abstract solutions ending with Π_0, $\langle \Pi_{n-1}, \ldots, \Pi_0 \rangle$, such that each Π_i is an i-th level abstract solution. Figure 10.6 shows a schematic diagram of this property.

By the upward-solution property, if there is no solution plan for a problem at an abstract level, then there is no solution at any lower level either. This fact follows directly from the upward solution property, since, if otherwise, a solution at any lower level would imply that solutions should exist at all higher levels, contradicting our initial assumption. This implication of the upward-solution property justifies our use of a top-down refinement strategy when planning with an abstraction hierarchy.

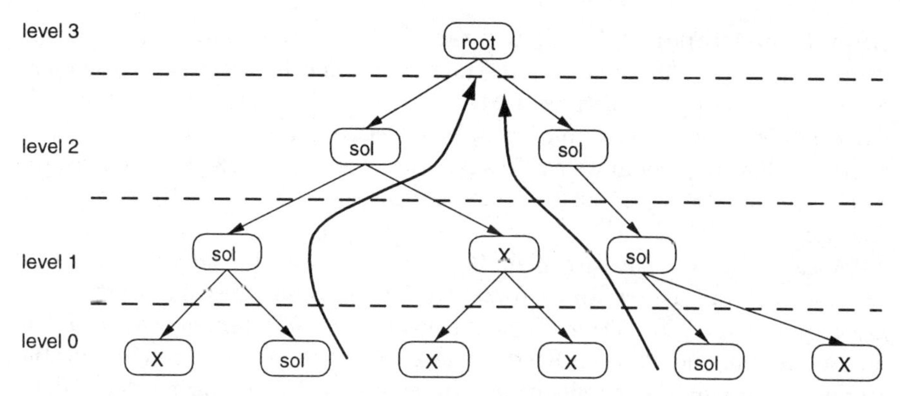

Fig. 10.6. Upward-solution property; in the figure every node marked **sol** is a solution node, and those marked with **X** are non-solution nodes.

Downward Solution Property. Now consider the existence property in a top-down direction. Suppose that, for a two-level hierarchy, an abstract solution exists. If this entails that a concrete-level solution also exists, then we say that the hierarchy satisfies the *downward solution property.*

In general, for an n-level abstraction hierarchy, we have the following definition. Let i be an integer ranging from 1 to $n-1$.

Definition 10.3.2 (Downward Solution Property). *Whenever any i-th-level abstract solution Π_i exists, there exists an $(i-1)$-th-level solution Π_{i-1}.*

By this definition, if an abstract solution Π_{n-1} exists, then there exists a sequence of abstract solutions ending with Π_0, $\{\Pi_{n-1}, \ldots, \Pi_0\}$, such that each Π_i is an i-th level solution.

By the downward solution property, if any solution is found at an abstract level, then we know immediately that a concrete-level solution exists for the original planning problem. Conversely, if there is no solution at the concrete level, then no solutions exist at any higher level either.

10.3.2 Refinement-Based Properties

In the definition of the above properties, one feature that is missing is the specification of a precise relationship between an abstract solution and a lower level solution. With upward-solution property, a concrete-level solution Π_c implies the existence of an abstract solution Π_a. But how is Π_c related to Π_a?

In this section, we specify a precise correspondence between solutions using a *refinement* relationship. By restricting the kind of refinement relations allowed, we could turn the above properties into useful search-control heuristics. In AI, transformations of this kind, where some abstract properties are turned into usable search control rules, are often referred to as *operationalization.*

Monotonic Property. Recall from the previous section that, with a precondition-elimination hierarchy, an abstract plan could be "refined" to a lower-level plan using the algorithms TOTALREFINE or PARTIALREFINE. Each algorithm is capable of returning several alternative refined plans for any given abstract plan. In general we can assume that there is a one-to-many relation Refine which, for any given abstract solution Π at level i, returns a set of solutions $\{\Pi_1, \Pi_2, \ldots, \Pi_B\}$ at level $i-1$.

Assume further that each plan Π_i in Refine(Π) leaves all causal links of Π intact. During the transition from Π to Π_i, no plan steps and causal links are removed from Π. Then Π_i is called a *monotonic refinement* of Π. In TOTALREFINE and PARTIALREFINE algorithms for precondition-elimination hierarchies, a monotonic refinement can be obtained by requiring that, when a new step is added to a plan, all threats with respect to a causal link that has been established at an abstract level are resolved. With a monotonic refinement, a plan never decreases in size as it evolves from one level to the next.

With the notion of monotonic refinement, we can now specify a variant of the upward-solution property. Let i be an integer between 0 and $n-2$.

Definition 10.3.3 (Monotonic Property). *Whenever an i-th-level solution Π_i exists, there exists an abstract solution Π_{i+1} at level $i+1$, such that Π_i is a monotonic refinement of Π_{i+1}. In other words, $\Pi_i \in$ Refine(Π_{i+1}).*

By this definition, if a concrete-level solution Π_0 exists then there exists a sequence of abstract solutions ending with Π_0, $\{\Pi_{n-1}, \ldots, \Pi_0\}$, such that each Π_i is an i-th level abstract solution, and every plan Π_{i-1} is a monotonic refinement of a previous plan Π_i.

If the monotonic property is established for an abstraction hierarchy, we could then use the following search heuristic during plan generation, with either a total-order planner or a partial-order planner: during refinement, if any abstract-level causal link is threatened, and if the threats cannot be resolved, then the plan can be discarded from the search space. This practice will not sacrifice any possible solutions.

With precondition-elimination hierarchies, upward-solution property always holds, implying that the monotonic property holds too. This result, first proven by Tenenberg [134] and then refined by Knoblock [80], can be derived by the following argument. Let Π_c be a concrete-level solution. We can replace each step in Π_c by its corresponding abstract-level step, simply by removing all level-0 preconditions. The resulting plan Π_c' is still a correct solution at the abstract level. Now, this plan is likely to contain a few redundant steps — steps that are not useful in achieving the final goal. If we remove these steps and their associated causal links, while maintaining the correctness of the plan, we could obtain a plan with three properties:

- the plan is still correct at the abstract level;
- the plan can be monotonically refined to Π_c; and

- the plan can be obtained by a planner, total-order or partial-order, at the abstract level.

By repeating this procedure for all higher levels, we could eventually obtain a sequence of plans, such that every plan in this sequence is a monotonic refinement of the previous plan.

Ordered Monotonic Property. With precondition-elimination hierarchies, the monotonic property is a universal property, in the sense that it is satisfied in every precondition-elimination hierarchy. When a property is universal, it is very likely that it has only weak heuristic power as a result of the well-known generality-effectiveness tradeoff. This is the case with the monotonic property; the monotonic property was implemented in an abstract planning system AbTweak [160]. It was found that in many cases, excessive backtracking is incurred due to violations of the abstract causal links. The end result is that for many hierarchies there is no observable improvement in planning efficiency over not using abstraction at all.

Similar observations led Knoblock to believe that it might be necessary to do away with backtracking across abstraction levels all together. This is enforced by a stronger property, known as the *ordered monotonic property*. Before explaining this property, we have to define what it means by an ordered refinement. A plan Π_i is an *ordered refinement* of an abstract plan Π if

- Π_i is a refinement of Π, and
- the new plan steps added into the abstract plan Π do not add or delete *any literals with a higher criticality value.*

In the simple-travel example, suppose that all city-level predicates are located at the abstract level, while the individual locations within a city are placed at a lower level. Then an ordered refinement of an inter-city plan can be obtained by restricting the intra-city movements to be within the boundary of a city.

Let i be an integer from 1 to $n-1$.

Definition 10.3.4 (Ordered Monotonic Property). *For every i-th-level solution Π, if Π has a refinement at level $i-1$, then* every *refinement of Π at level $i-1$ is an* ordered refinement *of Π.*

Obviously not every hierarchy satisfies the ordered monotonic property. Take the TOH domain for example. Consider a hierarchy where we assign the Onsmall and Onmed literals a higher criticality value than the Onlarge ones. And consider achieving a single goal literal

$$goal = \text{Onlarge(Peg3)}$$

Then no refinement of the abstract plan is an ordered refinement. To see this, observe that the abstract plan is in fact an empty plan. A refinement of the empty plan at level 0 is

$$\mathbf{s}_{init}$$
$$\downarrow$$
$$\texttt{moveS}(\text{Peg1}, \text{Peg3})$$
$$\downarrow$$
$$\texttt{moveM}(\text{Peg1}, \text{Peg2})$$
$$\downarrow$$
$$\texttt{moveS}(\text{Peg3}, \text{Peg2})$$
$$\downarrow$$
$$\texttt{moveL}(\text{Peg1}, \text{Peg3})$$
$$\downarrow$$
$$\mathbf{s}_{finish}$$

In this refinement, it is clear that in order to achieve the Onlarge(Peg3) goal, the abstract literals with predicates Onsmall and Onmed are all altered. Therefore, by the definition of ordered refinement, this hierarchy violates the ordered-monotonic property. Note that this conclusion holds for both total-order and partial-order refinements.

The above example also reveals one of the shortcomings of *ordered monotonic hierarchies.* The requirement that no low-level refinements alter any high-level literal is in fact too strong; a seemingly harmless addition to a plan could make property invalid. On the other hand, despite its over-restrictiveness, it still doesn't achieve one of its initial aim — to guarantee that there is no backtracking across abstraction levels.

Take the simple travel example. Assume that we place all In-City literals at level 1, and all At-Loc literals at level 0. This hierarchy is monotonically ordered, since a refinement at level 0 would only patch up the intra-city movements without venturing out of the parimeter of a city.

However, excessive backtracking between levels 0 and 1 can still occur. Any inter-city route that has been worked out at the abstract level could be found to be a dead end at level 0, due to unexpected road close-downs and traffic jams. In this case, re-planning for another inter-city route would then be necessary. In general, the ordered-monotonic property is neither necessary nor sufficient for ensuring that backtracking never occurs across abstraction levels.

Downward Refinement Property. In response to this issue, Bacchus and Yang [13] identified a property as the operationalized version of the *downward-solution property* — the *downward refinement property*, or the *DRP.*

Definition 10.3.5 (Downward Refinement Property). *Every abstract solution at level i can be* monotonically refined *to a solution at the next lower level, level $i-1$.*

The implication of the DRP is straightforward: if a hierarchy satisfies the DRP, then during planning, we only need to keep one copy of an abstract plan. From the DRP we know that this plan will lead to a correct plan at the next level. By induction, we also know that a concrete-level solution will eventually be found. Thus, the DRP could be used to dramatically prune a search space without loss of completeness.

Just like the ordered monotonic property, and unlike the monotonic property, the DRP is not universal; it is not satisfied by every precondition-elimination hierarchy. Nor is it satisfied by every task-network hierarchy, which we will explain in Chapter 12. What we would like to do next is to study the behavior of an abstract planning system as its hierarchy *approaches* this property, in a probabilistic sense. We can then state the impact of a *near-DRP* hierarchy on planning efficiency, which should have a wider range of application than the DRP in its strictest form.

10.4 An Analytical Model

We now analyze the benefits of near-DRP hierarchies, by studying a probabilistic model of abstract search. The search space explored by a hierarchical problem-solver can be viewed as a *tree* generated by the abstraction hierarchy (see Figure 10.7). In this tree each node at level i represents a complete i-th level abstract solution. The children of a node represent all of the different refinements of that solution at the next (lower) level of abstraction. The leaf nodes are complete concrete-level solutions. The task in searching through the space of abstract solutions is to find a path from the root down to a leaf node representing a correct concrete-level solution. Each node on the path must be a correct i-th level abstract solution to the problem at hand and must be a refinement of the $(i+1)$-th level abstract solution represented by its parent. That is, we are searching for an abstract solution that can be successfully refined through the levels of abstraction down to the concrete level.

The work in searching this tree comes from the work required to find the solution at each node. This solution is a refinement of the solution that has already been found at the node's parent. Hence, finding it simply involves

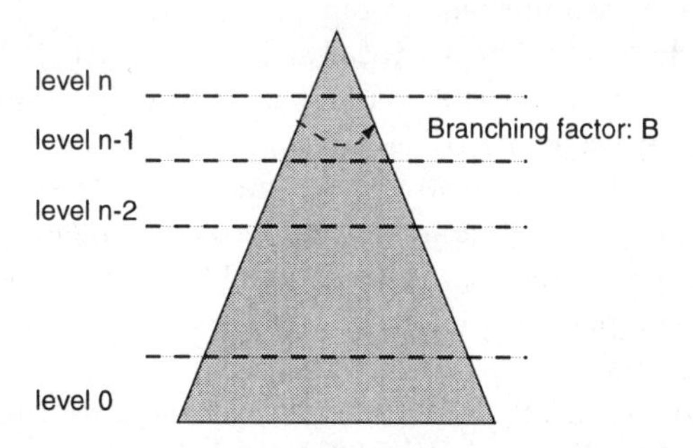

Fig. 10.7. Our analytical model: a tree of abstract solutions, with a branching factor B and a depth $n+1$.

solving all of the subproblems that arise when the parent's solution is moved down to this next lowest level. Since the number of subproblems increases as we move down the tree, the total amount of work will depend both on the number and depth of the nodes examined during search.

More specifically, the root of the tree represents a special length-one solution to every problem: a universal solution. Its presence is simply a technical convenience. The levels of the tree are numbered $\{n, \ldots, 0\}$ with the root being at level n and the leaves at level 0. Hence, discounting the universal solution at level n, our abstraction hierarchy has n levels.

The tree of abstract plans will have a branching factor that, in general, varies from node to node. This branching factor is the number of $(i-1)$-th level refinements possible for a given level-i solution, i.e., the number of children a node at level i has. Let the maximum of these branching factor be B. For simplicity we will use B as the branching factors for all nodes in the tree. Note that B has no straightforward relationship with the branching factor generated by the operators when searching for a specific solution, which we will denote by lower case b.

Under the assumption that the subproblems are independent, the number of refinements of a given i-th level solution, B, will be the product of the number of different solutions to each of the subproblems. Hence, it might be thought that B would grow exponentially as we move down the tree as the number of subproblems is growing exponentially. However, as we move down the levels of abstraction the subproblems become more and more constrained: at each level the solutions must preserve more and more conditions achieved at the higher levels of abstraction. Hence, the number of different solutions to the subproblems drops as we move down the tree. The exact balance between these two effects is difficult to determine, but we have found that our assumption of a constant branching factor B yields a model that is supported by empirical results.

10.4.1 Assumptions

At this stage for simplicity and ease of presentation we carry through our analysis under some assumptions. In our presentation, we don't restrict ourselves to total-order plans and to total-order refinements. We assume that each i-th level plan, when taken to a lower level to refine, will contain certain new subproblems. A schematic diagram for total-order refinement is shown in Figure 10.8, and one for partial-order refinement is shown in Figure 10.9. In precondition-elimination hierarchies, these subproblems arise due to the newly introduced preconditions at level $i-1$. For example, in the simple travel domain a new precondition might be a new location for the agent to arrive at, within a city.

The amount of work required to solve each subproblem individually is assumed to be $O(b^k)$. Here b is the *effective branching factor* of search. In forward-chaining, total-order planning b is simply the number of applicable

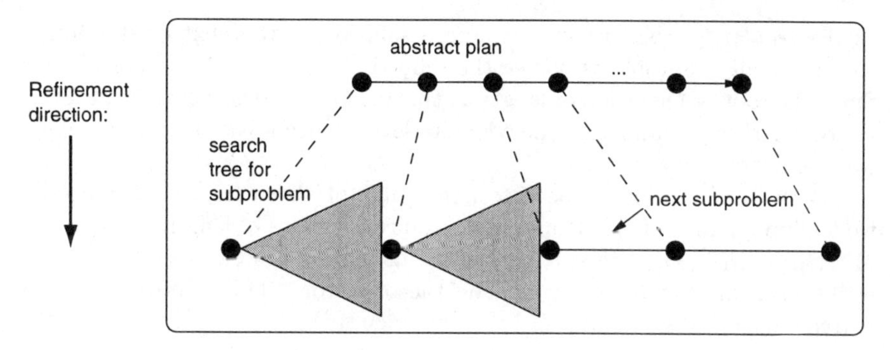

Fig. 10.8. Subproblems in a total-order refinement

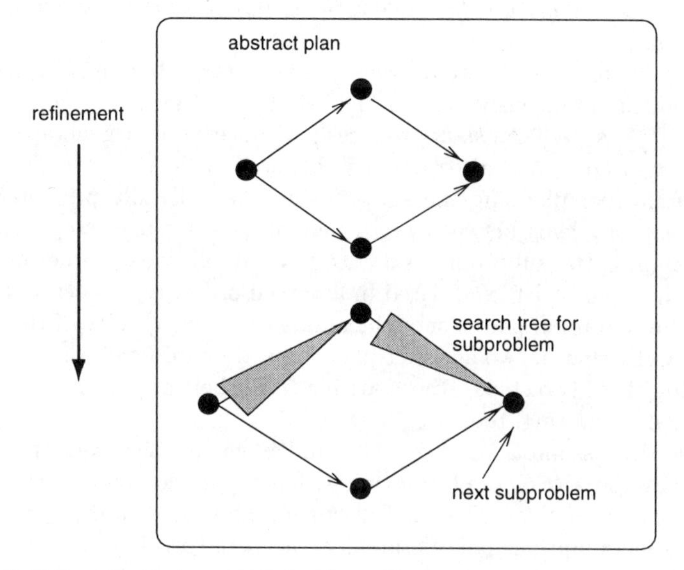

Fig. 10.9. Subproblems in a partial-order refinement

operators per state. In backward-chaining, partial-order planning, b is derived from equation (3.5):

$$B_b = ((B_{new} + B_{old}) * r^t)^P$$

where B_{new} is the number of new steps added in each search node, B_{old} is the number of existing plan steps used for achieving each precondition. P is the number of preconditions for each plan step, t the number of threats per node expansion, and r the number of ways to resolve each threat.

Our assumptions are as follows.

Regularity.
We assume that it takes approximately k new plan steps to solve every subproblem, where k is constant across abstraction levels. This is referred to

as the *regularity assumption.* Refining a solution to the next level amounts to solving all subproblems, hence the refined solution will be k times larger. Since the root is a solution of length 1, this means that the solutions at level i are of length k^{n-i}, and that the concrete-level solution is of length k^n, which we also denote by ℓ.

As this assumption degenerates, the value of abstraction degenerates. If we end up having subproblems which require solutions of length $O(\ell)$ instead of $O(k) = O(\ell^{1/n})$, then solving them will require search of $O(b^\ell)$ where b is the branching factor generated by the operators. This is no better than search without abstraction.

Independence Assumption.
We assume that the individual subproblems can be solved without significant interaction.

If, say, s subproblems interact we will have to search for a plan that solves all of them simultaneously. Such a plan would be of length $O(sk)$ and would require $O(b^{sk})$ search. As sk approaches ℓ we once again degenerate to search complexity of $O(b^\ell)$ and abstraction yields no benefits.

In total-order planning the subproblems are naturally presented in the form of *gap* problems between every pair of abstract plan steps. In partial order planning, two subproblems do not have to follow each other in time sequence; they can be left unordered in a partial-order. In this sense, the independence assumption places enough confidence in the quality of the abstraction hierarchy that, if two abstract plan steps are unordered, then the details concerning these two steps—those arising from their new preconditions—are largely non-interfering, too.

These two assumptions, regularity and independence, correspond to ignoring that the abstract solution might fail to provide an effective decomposition of the problem. The brief discussion above indicates, however, that when these assumptions fail, hierarchical problem-solving can quickly degenerate to being worse than non-hierarchical problem-solving. Hence, these two assumptions are required before we can obtain any interesting behavior from the abstraction hierarchy.

Existence of a Solution.
We assume that a concrete-level solution exists. This assumption is only needed to simplify our presentation. Actually, the analytical model we develop will also cover the case where this assumption fails.

Upward Solution Property.
We assume that the upward-solution property holds of the hierarchy. This property simply says that if a concrete-level solution Π_0 exists then there exists a sequence of abstract solutions ending with Π_0, $\{\Pi_n, \ldots, \Pi_0\}$, such that each Π_i is an i-th level abstract solution, and Π_{i-1} is a refinement of Π_i. That is, there is a sequence of refinements yielding a concrete-level plan.

Shortest Solution.
We assume that the abstract planner can find the shortest concrete-level solution for a problem. Let the solution length of the concrete-level optimal plan be l. By the above assumptions, $l = k^n$.

This assumption is also crucial in establishing the efficiency argument. If in the abstract space there are several alternative abstract plans, all with the same cost, then when we search in the abstract space we are blind as to which abstract plan to choose for refinement. In the worst case we could be led to an exponentially longer solution at the concrete level, leading to a degeneration of hierarchical problem-solving. This phenomenon was first pointed out by Backstrom and Jonsson [16].

Most of the above assumptions coincide with those considered by Knoblock [80], although they are not identical. Knoblock's analysis showed that with the above assumptions and when the DRP holds, using abstraction can yield an exponential-to-linear reduction in planning time. This result corresponds to the best-case behavior of an total-order abstract planning system. Below, we analyze the average-case search behavior for our general model.

10.4.2 The Probability of Refinement

If a hierarchy has the DRP then every solution at abstraction level i can be refined to a solution at abstraction level $i-1$. This means that once we have found a path down to level i, whose terminal node represents a correct i-th level solution, we are assured that this node has a child representing a valid $i-1$-th level solution. That is, we are assured that the node can be refined to the next lower level. Hence, we need never reconsider the initial part of the path, we just need to extend it until we reach a leaf: there is no backtracking across abstraction levels.

A reasonable way to examine the behavior of hierarchies in which the DRP fails is to assign a probability, p, to the event that a given i-th level solution can be refined to level $i-1$. The DRP now corresponds to the case $p = 1$. When $p < 1$, however, we might build a path of correct solutions from the root down to a node at level i, and then find that this node is not refinable to the next level. This will force a backtrack to the penultimate node at level $i+1$ to find an alternate level i solution, one which is refinable. This may cause further backtrack to level $i+2$, or search may progress to lower levels before backtracking occurs again.

Since we are assuming that a concrete-level solution exists, and that the upward solution property holds, we know that there is at least one path of correct solutions in the tree from the root to a leaf node. Hence, although the worst case will require an exhaustive search of the tree, search will eventually succeed in finding a good path.

What we wish to accomplish, then, is to determine the average case complexity of search in abstraction hierarchies in which (1) the probability that a given node in the abstraction search tree can be refined is p, and (2) there is at least one *good path*, i.e., a path of good nodes, from the root to a leaf in the tree.

If a node is refinable to the next level, it will have B children, under our assumption that the number of refinements be treated as a constant. Each of these children is itself refinable with probability p. During search if we encounter a node that is refinable we would have to examine all of the subtrees headed up by its B children before we can conclude that it is a dead end. On the other hand, if the node is not refinable we can backtrack right away. Hence, as far as search is concerned, a node that is refinable has B children, while a node that is not refinable has no children.

Analytically, however, we can find the average case complexity of searching such trees by considering randomly labeled *complete* trees. A complete abstraction tree is simply a tree in which *every* node, refinable or not, has B children and which has height $n+1$. Hence, it has

$$N = \frac{B^{n+1} - 1}{B - 1}$$

nodes. We randomly label each node in this tree as being refinable (*good*) with probability p, or not refinable (*bad*) with probability $1 - p$. Each of the 2^N distinct trees that can be generated by this process has probability $p^g(1-p)^{N-g}$, where g is the number of good nodes in that tree. Now the average case complexity for searching this collection of trees is simply the sum of the search effort required for each tree times the probability of that tree. The average case complexity for searching all of these trees is the same as the average case complexity of searching the original set of trees in which a bad node has no children. Each original tree in which bad nodes have no children represents a set of complete trees: each one generated by a different labeling of the subtrees under the bad nodes. The probability of the original tree is exactly the same as the sum of the probabilities of the complete trees in this set.

We wish to restrict our attention to the case where each tree contains at least one good path. That is, we assume that a solution exists at the concrete level. When we generate the set of differently labeled complete trees, some of them will not contain a good path from the root to a leaf. We remove these trees, and re-normalize the probabilities of the remaining trees so that they sum to 1. That is, we take only conditional probabilities.

10.4.3 Analytical Result

Let $\mathsf{GoodTreeWork}(i)$ be the expected amount of computation required to search a *good subtree* with root at level i, i.e., a subtree which contains at least one good path from its root to a leaf. To examine a good tree we have to expand its root node. Then we must search the subtrees under the root,

Table 10.8. Asymptotic search complexity for different regions when a solution exists

p	$(0, 1/B)$	$1/B$	$(1/B, 1)$	1
Variable n	$O(\ell)$	$O(\ell \log(\ell))$	$O(\ell (pB)^{\log_k(\ell)})$	$O(\ell)$
Constant n	$O(\ell b^{\sqrt[n]{\ell}})$	$O(\ell b^{\sqrt[n]{\ell}})$	$O(\ell b^{\sqrt[n]{\ell}})$	$O(\ell b^{\sqrt[n]{\ell}})$

looking for a good subtree rooted at the next level. Once we find such a subtree we never need backtrack.

The average-case time complexity of abstract planning is $\mathsf{GoodTreeWork}(n)$, which can be calculated by the following theorem.

Theorem 10.4.1

$$\mathsf{GoodTreeWork}(n) = \begin{cases} O(b^k k^{n-1}) & p = 1. \\ O(b^k k^{n-1}) & p < 1/B \\ O(b^k k^{n-1} n) & p = 1/B, \\ O(b^k k^{n-1} (pB)^{n-2}) & 1/B < p \leq 1 \end{cases} \tag{10.1}$$

The proof requires familiarity with the mathematics of *branching processes* [9]. Interested readers may find a thorough proof in [13].

With the result from this theorem, we can now discuss average-case search complexity in different regions of refinement probability. There are two cases to consider: hierarchies with a constant number of levels and those with a variable number. In certain domains we can make n, the number of abstraction levels, vary with ℓ, the length of the concrete solution. For example, in the Towers of Hanoi domain we can place each disk at a separate level of abstraction [80]. In other domains, e.g., blocks world, it is not so easy to construct a variable number of abstraction levels, and n is generally fixed over different problem instances.

Since each refinement multiplies the length of the abstract solution by k, we have that the concrete solution will be of size k^n, i.e., $k^n = \ell$. We want to express our results in terms of ℓ. If we can vary n with ℓ then we can ensure that k remains constant and we have that $n = \log_k(\ell)$. In this case, b^k will become a constant. Otherwise, if n is constant, $k = \sqrt[n]{\ell}$ will grow slowly with ℓ. In this case, $b^k = b^{\sqrt[n]{\ell}}$ grows exponentially with ℓ, albeit much more slowly than b^ℓ (cf. [80]). This essential difference results in different asymptotic behavior for the two cases, n variable and n constant. Table 10.8 gives the results of our analysis for these two cases expressed in terms of the length of solution ℓ.

We conclude from this table the following key points:

1. Non-hierarchical search requires $O(b^\ell)$; hence, it is evident from the table that when $0 \leq p \leq 1/B$ and when $p = 1$ abstraction has a significant benefit. Our result for $p = 1$ agrees with that of Knoblock [80]: here

we have the DRP and all of his assumptions hold. Our results for the region $0 \leq p \leq 1/B$, extend his analysis, and indicate that abstraction is theoretically useful when the probability of refinement is very low.

2. The region $1/B < p < 1$ corresponds to the worst region for search complexity. For variable n, in this region we increase by a factor, $O((pB)^n)$, that is exponential in n. In these regions it is not always advantageous to increase the number of abstraction levels n, especially in the region $1/B < p < 1$. As p increases in this region search first becomes harder and then easier. This can be seen from equation (10.1).
   Intuitively, what is occurring in this region is that the bad subtrees are becoming increasingly difficult to search. We have to search a larger and larger proportion of a bad subtree before we can recognize that it is bad. At the same time, as p grows, the number of bad subtrees we have to search is decreasing. These two trends fight each other, with the total work required first growing and then shrinking, until we reach $p = 1$ where the number of bad subtrees we have to search falls to 0.

Our analysis also tells us that if the number of possible refinements for an abstract solution, B, is large, then searching the abstraction tree is more expensive in the worst region $1/B < p < 1$. This is to be expected: the abstraction tree is bushier and in this region we have to search a significant proportion of it. Also of interest is that B does not play much of a role outside of this region, except, of course, that it determines the size of the region. Hence, if we know that the DRP holds or if the probability of refinement is very low, we do not have to worry much about the shape of the abstraction tree. However, without such assurances it is advantageous to choose abstraction hierarchies where abstract solutions generate fewer refinements. For example, this might determine the choice of one criticality ordering over an alternate one in precondition-elimination style abstraction.

Another interesting result of our analysis is that search is most complex in the middle region. When the probability is low that an abstract solution can be refined, more abstract solutions need to be examined before a refinable one is found. However, not much work is required to detect unrefinability. On the other hand, when the probability of refinability is high, not too many abstract solutions need be checked before a good one is found. The worst case is in the middle. There a significant fraction of the abstract solutions are unrefinable, and it can take a great deal of work to discover that they are unrefinable. The existence of such a phase boundary agrees with empirical studies of Cheeseman et. al. [30], who found that the hard cases of many problems tend to cluster in the phase boundary between very many solutions and very few solutions.

10.5 Open Problems

Following the aforementioned results in abstraction, a few critiques emerged. With these, the study of abstraction, like any other scientific discipline, is likely to undergo another cycle of evolution.

Of the critiques, the first is by Lansky and Getoor [87], who argued that many of the formal properties, such as the ordered-monotonic property, are in fact heavily dependent on a lack of subgoal-interactions between levels of hierarchies. This strict condition is often the reason to collapse an entire hierarchy into a single level, thereby diminishing the value of abstraction altogether. What is needed is a hierarchy that can tolerate a certain number of interactions between levels, while keeping the computational complexities well under control.

Another critique is offered by Smith and Peot [123]. They demonstrated that the ordered-monotonic property is in fact not able to handle interactions arising from resource allocations. As we will see in the next chapter, part of their criticism can be dealt with under a property we call near-DRP. But the issue still largely remains; it is likely that some other unknown properties might exist that can characterize a different type of interactions, such as resource contention.

A third criticism is made by Bergmann and Wilke [20]. They argued that in order for an abstraction planner to be truly efficient, it is crucial to provide an *explicit* refinement function for plans between successive levels in a hierarchy, and that at different levels, separate planning languages are used. In their work these refinement functions are provided by the user, and their empirical results have shown that having an explicit refinement function could make the planner dramatically more efficient than one with the precondition-elimination type of abstraction.

Finally, Backstrom and Jonsson [16] have provided an analysis of hierarchical planning where the shortest-solution assumption is removed. They showed that in this case using abstraction it is possible to obtain an exponentially more costly solution plan (see Section 11.5 for a more detailed discussion).

10.6 Summary

In this chapter we provided a framework in which to study abstraction. We described the precondition-elimination type of abstraction as a case study, and outlined a number of properties with which one could judge the quality of an abstraction hierarchy. Finally, we performed an average-case analysis of the value of abstraction, showing that, under a number of assumptions, the refinement probability of a hierarchical planner largely determines the efficiency of abstract problem solving.

10.7 Background

Korf gave an early account of the value of abstraction in search [83]. Several books on hierarchical problem solving are also available. Knoblock's book [80] presents a comprehensive coverage, both theoretical and empirical, of the ALPINE system. Chapter 4 of the book by Allen, Kautz, Pelavin, and Tenenberg [6] discusses inheritance hierarchies and various properties associated with precondition-elimination hierarchies. Wilkins' book [147] discusses various aspects related to hierarchical task network hierarchies in the context of the SIPE planner. Sacerdoti's book [114] makes good reading on the historical background in this topic. Holte et al. [67] investigated hierarchical A*, a forward-chaining, total-order planner using abstraction. Hierarchical problem solving is also well explored in theorem proving. See [58] for a good introduction.

The material in this chapter came from several sources. Sections 10.1 and 10.2 are adapted from a Computational Intelligence article with Tenenberg and Woods [160]. Section 10.3 is adapted from articles appearing in AAAI90 (with Tenenberg) [159], AAAI91 (with Knoblock and Tenenberg) [79], AAAI92, and IJCAI91 (with Bacchus) [12, 11]. Section 10.4 is adapted from an Artificial Intelligence journal article with Bacchus [13].

11. Generating Abstraction Hierarchies

Having established a theoretical understanding of when hierarchical planning can improve planning efficiency, we now discuss how to ensure the *downward refinement property*, or the DRP. Recall that the DRP holds for a hierarchy if every abstract solution can be refined to a concrete solution, given that a concrete solution exists. In this chapter, we concentrate on precondition-elimination hierarchies, for which we provide a useful *syntactic* condition that is sufficient to guarantee the DRP. Hence, in certain cases we can detect if a hierarchy is "good" in the sense that, under the assumption of easy and independent intermediate goals, we know, via the analytical models, that such hierarchies yield a significant speed-up over non-hierarchical planning.

With the syntactic condition in place, we also design a hierarchy generator called HIPOINT. This system builds on the ALPINE system of Knoblock [80], which is an automatic hierarchy generator for precondition-elimination hierarchies. HIPOINT takes the hierarchy suggested by ALPINE and improves it using information gathered during a testing phase. These improvements are based on the results of our analysis as well as the syntactic condition we develop. Our empirical results demonstrate that HIPOINT can generate hierarchies that offer a significant improvement in performance over non-hierarchical planning.

11.1 Syntactic Connectivity Conditions

To know whether the DRP holds for a given precondition-elimination hierarchy, one way is to check, for each pair of operators that could possibly appear in sequence in a plan, whether it is possible to "connect" the lower level instantiations of the operators by a plan segment. We hope that this plan segment will leave all upper-level causal links intact, thus producing a monotonic refinement.

Clearly, it is computationally intractable to check for the existence of plan segments of any arbitrary length. So, instead we realize our test for this condition by checking for solutions of length k. This means that we only need $O(b^k)$ work to perform this test for each pair of operators. We keep k fixed and small so that the test can be run efficiently. An alternate way of realizing the test is to simply search for a solution under a fixed time bound.

Both of these approaches mean that the syntactic test is only a sufficient, not necessary test for the DRP.

Suppose that we are given a precondition-elimination hierarchy with n levels. Assume that level 0 corresponds to the concrete level. The *k-ary connectivity test* is run on a pair of operators α_1 and α_2 and is also dependent on the level of abstraction i. The test is in the form of an implication. Hence, a pair of operators will pass the test if they fail to satisfy the first condition, i.e., the antecedent. To describe this condition, we first introduce a notion for applying an operator to a set of literals. Let α be an operator, and L a set of literals. $\alpha(L)$ is a set of literals obtained from L, by removing all literals in L which are not consistent with an effect of α. That is,

$$\alpha(L) = L - \{l \mid l \in L \text{ and } \neg l \in \mathsf{Eff}(\alpha)\}$$

For the antecedent test the conditions are

Definition 11.1.1 (k-ary Antecedent Test). *Two operators α_1 and α_2 satisfy the k-ary antecedent condition if*

1. $\mathsf{Abs}(i, \alpha_1(\mathsf{Pre}(\alpha_1))) \cup \mathsf{Abs}(i, \mathsf{Pre}(\alpha_2))$ *does not contain both a literal and its negation; and*
2. $\mathsf{Abs}(i-1, \mathsf{Pre}(\alpha_2)) \neq \mathsf{Abs}(i, \mathsf{Pre}(\alpha_2))$.

By the first condition α_1 and α_2 are operators that could appear in sequence in an i-th level plan, and by the second, α_2 will create a subproblem, so that there is the possibility of a problem when refining to this level.

If α_1 and α_2 fail this condition, the k-ary connectivity test returns success; this pair of operators will not cause a problem during refinement. If the antecedent test is satisfied, the operator pair must then satisfy the consequent test, which is as follows.

Definition 11.1.2 (k-ary Connectivity Test). *Given a pair of operators α_1 and α_2 that satisfy the k-ary antecedent test, there must exist a sequence of k, $i-1$-level abstract operators $\beta_1, \ldots, \beta_k$ such that:*

1. *None of the β_i adds or deletes literals with criticality higher than $i-1$, and*
2. *The sequence of operators β_i is a solution to the problem*

$$\langle \mathsf{Abs}(i-1, \alpha_1(\mathsf{Pre}(\alpha_1))), \mathsf{Abs}(i-1, \mathsf{Pre}(\alpha_2)) \rangle.$$

If the consequent test succeeds, then we return success, otherwise failure.

We can generalize the test to all operators in the domain:

Definition 11.1.3 (k-ary Necessary Connectivity). *Let $\mathcal{O}_{\geq i}$ be the set of operators whose effect lists contain at least one literal with criticality value greater than or equal to i. If every pair of operators α_1, α_2 from $\mathcal{O}_{\geq i}$ passes the* k-ary connectivity test, *then levels i and $-$ are k-ary connected. If every pair of levels i, $-$ in the hierarchy is k-ary connected then we say that the hierarchy satisfies the condition of k-ary necessary connectivity.*

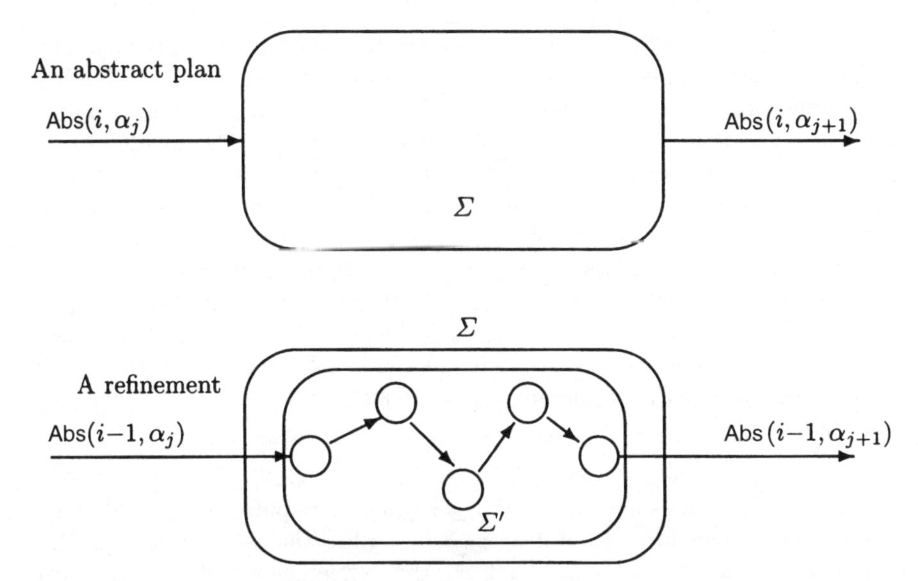

Fig. 11.1. A plan segment and its refinement

Intuitively, k-ary necessary connectivity is saying the following. Since any partial-order plan can be considered as sequences of total-order plans, we can concentrate on the most elementary component of a total-order plan—a pair of operators. For every pair of operators, if they might be sequenced in a plan at level i (the antecedent test), there must exist a sequence of $i-1$ abstract operators β_i that can correctly solve any subproblem that might result from the refinement of α_1 and α_2 to the next lower level, $i-1$ (the consequent test; see Figure 11.1). Any subproblem at abstraction level $i-1$ would have to start from a state that must include all the *weakest* conditions established as a result of applying α_1. This state is specified by the application of α_1 to its own preconditions, i.e., $\alpha_1(\mathsf{Pre}(\alpha_1))$, and we simply have to take its $i-1$ abstraction. Similarly, the subproblem would have to end in a state that achieves the $i-1$ preconditions of operators α_2. This argument can easily be formalized to yield:

Theorem 11.1.1 *k-ary necessary connectivity is sufficient to guarantee the DRP.*

As we increase the parameter k we increase the complexity of performing this test. The complexity increases exponentially with k as we are searching through the space of operator sequences of length k.

Although we can keep each individual run of the k-ary connectivity test at a reasonable level of complexity, by keeping k small, we still have to test every pair of instantiated operators to determine necessary connectivity. This might be quite a large number, although various techniques can be applied to minimize the number of tests we need. For example, the operator variables are subject to type constraints so not all instantiations are legal; similarly, if two constants have exactly the same properties we do not have to use both in our instantiations. However, we will not delve into such details here. Our main use of necessary connectivity will be to estimate refinement probabilities by subjecting random pairs of operators to the connectivity test.

In practice, however, it may be too strong to require that every abstract plan have a monotonic refinement. In that case, we can relax this requirement and consider instead hierarchies that are close to having the DRP. In particular, we can say that a hierarchy is *near-DRP* if it satisfies the following conditions:

1. Let the refinement probability for level i be p_i, $i = 0, 1, \ldots, n$. Let θ be a user-defined threshold, where $0.5 \leq \theta \leq 1$. We require that $p_i \geq \theta$. That is, the refinement probabilities must be no less than a given threshold.
2. $p_j \geq p_i$ for all $j > i$. That is, the refinement probabilities must be monotonically increasing as we move down the levels of abstraction.

Both of these conditions are motivated by our analysis in the last chapter. As discussed in Chapter 10 sufficiently high refinement probabilities will result in less search, and typically increasing refinement probabilities are more effective.

11.2 HIPOINT

A good hierarchy should have the ability to avoid interactions with higher-level achievements, and it should ensure that for every abstract plan a low-level refinement exists with high probability. In this section we present an algorithm, HIPOINT, that automatically constructs abstraction hierarchies possessing both of these properties, namely, the ordered monotonic property (OM) and the near-DRP. We would like to build on hierarchies with the OM property, because, due to its restriction that no low-level refinements can change a high-level literal, a hierarchical planner need not check on any of its abstract-level causal links to see if they are violated. The OM hierarchy guarantees that they will not. Thus, considerable computational gain can be achieved by using an OM hierarchy.

The algorithm HIPOINT is presented in Table 11.1. Informally, HIPOINT takes as input a set of operators, initial states, and goal states of a domain. First it invokes ALPINE [80] to generate a partially-ordered graph that represents a set of hierarchies with the ordered monotonic property (OM) for

Table 11.1. The HIPOINT algorithm for creating a hierarchy

Input: A set of operators $\mathcal{O}$, initial states *init* and goal states *goal*.
Output: A criticality assignment to the predicates, such that the abstraction hierarchy satisfies the OM property and is close to being near-DRP.

Algorithm HIPOINT(**O**,*init*,*goal*)
1. *graph* := ALPINE($\mathcal{O}$, *init*, *goal*);
2. **for** every pair of nodes n_i and n_j in *graph*, such that
3. there is no path from n_j to n_i in *graph*, **do**
4. $prob(n_i, n_j)$:= FIND-PROBABILITY($n_i, n_j, \mathcal{O}$)
5. **end for**
6. *graph* := COLLAPSE-NODES(*graph*, *prob*);
7. *hierarchy* := AUGMENTED-TOP-SORT(*graph*, *prob*);
8. **return**(hierarchy);

---

the given domain. Recall that the property states that with the abstraction hierarchy, every refinement leaves all literals in the higher levels intact.

Each node n_i in the graph represents a set of predicates that should be assigned the same criticality in an abstraction hierarchy. An arc from a node n_i to n_j denotes that plans for achieving subgoals whose predicates are in the set n_j will not affect any predicates in the set n_i. That is, if we have an abstract plan involving the achievement of literals whose predicates are in n_i, then we can refine that plan to achieve literals in n_j without affecting any of the n_i literals. The graph has the property that every total order of the nodes that extends the partial order, represents a hierarchy with the Ordered Monotonic Property.

Next, steps 2–5 of the algorithm assign an estimated refinement probability to every pair of nodes n_i and n_j such that $crit(n_i) > crit(n_j)$ is allowed by the partial order defined by the graph. Step 6 processes the nodes in the graph using the additional information provided by the estimated refinement probabilities. The procedure AUGMENTED-TOP-SORT returns a criticality assignment to each predicate, such that the resultant hierarchy has the OM property and is close to being near-DRP.

We cannot guarantee near-DRP for two reasons. First, we are only able to obtain *estimates* of the refinement probabilities, so the true refinement probabilities might fail to satisfy the near-DRP property. Second, the OM property forces the placement of some literals above others. Since it does not consider refinement probabilities, the placement of literals forced by OM might result in the violation of near-DRP. HIPOINT attempts to find the best hierarchy, with respect to being near-DRP, among those hierarchies that satisfy the OM property. The hierarchy is then returned as the output of the algorithm.

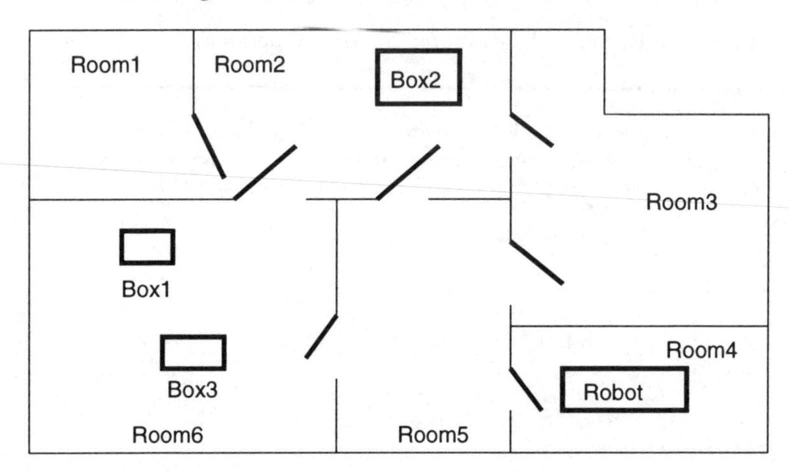

Fig. 11.2. Robot-box planning domain

Below, we explain in detail each of the major components of HIPOINT, with the help of an extension of the simple robot domain. In this domain there is a robot and a number of connected rooms between which the robot can move. Between any two rooms there may be a door, which can be open or closed. In addition, there are also a number of boxes, which the robot can either pull or carry from one location to another. Figure 11.2 shows one configuration of the domain. The operators in this domain are presented in Table 11.2. There is one additional operator not shown in the table, `carry-thru-door`(?b, ?d, ?r2, ?r2); this operator is identical to `pull-thru-door` except that the box, ?b, must be Loaded instead of Attached.

Our representation for this domain includes the following predicates: Box-Inroom(?b,?r) representing that box ?b is in room ?r; Attached(?b) representing that box ?b is attached to the robot; Loaded(?b) representing that box ?b is loaded onto the robot, Open(?d) representing that door ?d is open. In addition, there are also a number of type predicates: Isdoor(Door), Openable(Door), Connects($Door_{12}$, R_1, R_2).

11.2.1 ALPINE

ALPINE is an abstraction-hierarchy generation algorithm designed and implemented by Knoblock [80]. Given the operator definitions for a given domain, ALPINE constructs a partially ordered graph of the literals. Each node in the graph denotes a set of literals that are to be assigned the same criticality value in the final hierarchy. If a node n_i precedes n_j in the graph, i.e., if there is a path from n_i to n_j in the graph, then we cannot place n_j above n_i in the final hierarchy; the criticality value of n_i must be greater than or equal to the criticality value of n_j. The algorithm ensures that every total

Table 11.2. Operators for the robot-box domain

Preconditions	*Effects*
pull-thru-door(?b,?d,?r1,?r2)	
Isdoor(?d), Isbox(?b),	Box-Inroom(?b,?r2), ¬Box-Inroom(?b,?r1)
attach-box(?b)	
Isroom(?r1), Isroom(?r2), Connects(?d,?r1,?r2), Attached(?b), Box-Inroom(?b,?r1), Open(?d)	
Isbox(?b), ¬Attached(?b)	Attached(?b)
load-box(?b)	
Isbox(?b), ¬Loaded(?b)	Loaded(?b)
open-door(?d)	
Isdoor(?d), Openable(?d), ¬ Open(?d)	Open(?d)

order supported by the graph will yield a hierarchy that has the ordered-monotonic property, whereby every refinement of an abstract plan leaves all the higher-level literals unchanged.

The core of the ALPINE algorithm is a syntactic restriction formalized in [79]:

Definition 11.2.1 (Ordered Restriction). *Let $\mathcal{O}$ be the set of operators in a domain. Let P_α be the preconditions of α that can be either added or deleted by some operator. Then a criticality assignment satisfies the ordered restriction if $\forall \alpha \in \mathcal{O}$, $\forall p \in P_\alpha$, and $\forall e_1, e_2 \in \mathrm{Eff}(\alpha)$,*

(1) $crit(e_1) = crit(e_2)$, and
(2) $crit(e_1) \geq crit(p)$.

That is, all the effects of an operator are required to have the same criticality, and that criticality must be at least as great as the operator's changeable preconditions. It was shown in [79] that if the assignment of criticality values satisfies this restriction then the hierarchy will satisfy the OM property.

The ALPINE algorithm implements this restriction, generating a graph representing partially ordered collections of literals. The ALPINE system then uses this graph to compute a particular total order and resulting hierarchy, using a collection of heuristics to pick the total ordering. The algorithm depends only on the operator definitions, not on the goals and initial situations. However, Knoblock has shown how to modify this algorithm so that it can generate problem-specific hierarchies that take into account a particular goal and initial state. This often results in a finer grained hierarchy [80].

We illustrate the algorithm via the robot-box example. When applied to the operators in this domain, ALPINE generates the graph shown in Figure 11.3. It is clear that there are six possible total orders that could result

Table 11.3. Robot-box domain hierarchy generated by ALPINE

Criticality	Predicates
4	Isdoor, Openable, Connects, Isbox, Isroom
3	Box-Inroom
2	Attached
1	Loaded
0	Open

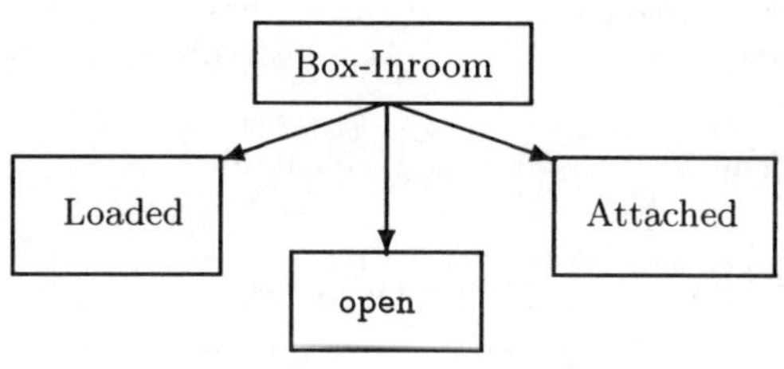

Fig. 11.3. Robot-box domain graph generated by ALPINE

from this graph, corresponding to the predicate Box-Inroom followed by permutations of open, Attached, Loaded.[1] Each total order corresponds to a different hierarchy. For example, for the total order,

$$\text{Box-Inroom} \prec \text{Attached} \prec \text{Loaded} \prec \text{Open}$$

the corresponding hierarchy is shown in Table 11.3.

11.2.2 Probability Estimates

If there is a path from n_i to n_j in the ALPINE graph, we cannot place the literals in n_j at a higher level of abstraction than the literals in n_i, without violating the OM property. However, for all pairs n_i and n_j that do not have this constraint we have a choice; we can, if we wish, place n_i above n_j in the final hierarchy. The next step of the HIPOINT algorithm is to determine the merit of such a placement. It does this by estimating the refinement probability that would exist between the levels n_i and n_j if n_i was in fact placed above n_j. If node n_i was placed before n_j, then all of the predicates in n_i would be placed at a higher level of abstraction than the predicates in n_j.

For all pairs of nodes n_i and n_j such that it is possible to place n_i above n_j, HIPOINT calls FIND-PROBABILITY(n_i, n_j) to estimate the probability of refinement going from predicates in n_i to predicates in n_j.

[1] The type predicates, like Openable, are not shown on this graph. Since these predicates are not affected by any of the operators, they are always placed at the highest level by ALPINE.

In its operation FIND-PROBABILITY first locates sets of operators O_i and O_j that achieve literals in n_i and n_j respectively. It generates its estimate of the refinement probability from n_i to n_j by determining how often pairs of operators in O_i pass the connectivity test, as in Definition 11.1.2, where we consider the predicates in n_i to be at level i and the predicates in n_j to be at level $i-1$.

Consider a plan at level n_i. Such a plan will typically have to achieve literals in n_i, hence it will contain operators from O_i. To refine this plan to the level of n_j, we will have to solve the subproblems generated by pairs of operators from O_i using operators from O_j. According to Definition 11.1.2 this means testing how often the planning problem

$$\langle \mathsf{Abs}(i-1, \alpha_1(\mathsf{Pre}(\alpha_1))), \mathsf{Abs}(i-1, \mathsf{Pre}(\alpha_2))\rangle$$

can be solved using operators that affect only literals at level $i-1$ and below, where α_1 and α_2 are members of O_i. Since we are ignoring all other nodes, we ignore all preconditions of α_1 and α_2 that are outside of n_i and n_j, except for type preconditions which are always at the top of the hierarchy.

For operators in O_i, GENERATE-RANDOM-PROBS chooses at random a pair of instantiated operators that passes the antecedent condition of the connectivity test. It then computes the problem shown above, which is specified by the consequent condition of the connectivity test. A whole collection of such pairs of operators are generated, and their resulting problems are returned as the set "random-probs".

FIND-PROBABILITY then calls a planner PLANNER, which tries to solve these problems. The PLANNER algorithm could be implemented as either a forward-chaining, total-order planner, or a backward-chaining, partial-order planner. In line 5 of the FIND-PROBABILITY algorithm, the PLANNER parameter O_j/O_i specifies the operators that can be used to solve the problem. This is the set of operators O_j with those operators that affect literals contained in n_i removed, i.e., all operators also in O_i have been removed using a set-difference operation. Each problem is solved under a fixed solution-length bound, this implements the "k-ary" limit part of the connectivity test. FIND-PROBABILITY accumulates a count of the number of times a solution is found.

The frequency of success serves as an estimate of the refinement probability, as we know that if two operators pass the test then the gap subproblem they generate during refinement can be solved. If GENERATE-RANDOM-PROBS fails to find any random problems, then it has failed to find any operators with level n_j preconditions among the set of operators that can be sequenced. Hence, it is unlikely that any difficulties will be encountered during refinement, and we estimate the refinement probabilities as being 1.

We now illustrate the procedure using our robot domain. Let n_i be {Box-Inroom}, and n_j be {Open}. The operator sets corresponding to the nodes are

Table 11.4. Algorithm FIND-PROBABILITY

Input: Operators O, predicate sets n_i, n_j.
Output: A refinement probability value.
Procedure FIND-PROBABILITY(n_i, n_j, O)
1. O_i := *Find-Operators*(O, n_i);
2. O_j := *Find-Operators*(O, n_j);
3. random-probs := GENERATE-RANDOM-PROBS(O_i, n_i, n_j);
4. **for** every random problem $\langle initial, goal \rangle$ in random-probs **do**
5.   **if** PLANNER$(initial, goal, O_j/O_i,$ solution-length-bound$) = Success$ **then**
6.   success-count := success-count + 1
7.   **end if**
8. **end for**
9. **if** |random-probs| = 0 **then**
10.   prob := 1
11. **else** prob := success-count/|random-probs|
12. **end if**
13. **return**(prob);

$$O_i = \{\texttt{pull-thru-door}, \texttt{carry-thru-door}\}$$

and $O_j = \{\texttt{open-door}\}$. Both of the operators in O_i can be sequenced in any manner as long as the room the first operator takes us into is identical to the room the second operator moves us out of; this constraint is enforced at the n_i level via the Box-Inroom precondition of these operators. Additionally, we ignore the precondition of Attached (or, Loaded in the case of `carry-thru-door`) as this precondition is in node other than n_i or n_j and it is not a type predicate.

So of those operators that can be sequenced, i.e., that pass the antecedent test, we will generate random problems that are instantiations of the following template arising from the connectivity test:

$$\left\langle \begin{array}{l} \text{Isdoor}(?d1), \text{Isbox}(?b) \\ \text{Isroom}(?r1), \text{Isroom}(?r2) \\ \text{Connects}(?d1, ?r1, ?r2) \\ \text{Box-Inroom}(?b, ?r2) \\ \text{Open}(?d1) \end{array} , \begin{array}{l} \text{Isdoor}(?d2), \text{Isbox}(?b) \\ \text{Isroom}(?r2), \text{Isroom}(?r3) \\ \text{Connects}(?d2, ?r2, ?r3) \\ \text{Box-Inroom}(?b, ?r2) \\ \text{Open}(?d2) \end{array} \right\rangle .$$

That is, the problem is to achieve the preconditions of the second operator in the state that results from applying the first operator to its precondition set, as specified by the connectivity test, for various instantiations of the operators. It is easy to see that this problem reduces to achieving `open`(?d2) for different instantiations of ?d2. A door can be opened with the operator `open-door`(?d) without affecting any higher level literals, as long as it is Openable. So we see that the number of solvable random problems will correspond approximately to the proportion of doors that are Openable in the

Table 11.5. A matrix of refinement probabilities

Predicates	Box-Inroom	Open	Attached	Loaded
Box-Inroom	0	0.5	1	1
Open	0	0	1	1
Attached	0	1	0	1
Loaded	0	1	1	0

domain. Hence, the refinement probability returned by FIND-PROBABILITY will depend on the probability of a door being openable. This agrees with intuition. If we place Open at a lower level of abstraction, the abstract level will be free to develop a plan ignoring the status of the connecting doors. If most doors can be opened this will generally not be a problem, but if most doors cannot be opened, most of the routes chosen at the abstract level will fail. That is, most of the abstract plans will not be refineable.

The result of running FIND-PROBABILITY on all eligible pairs of nodes is a matrix of estimated refinement probabilities. In one of our tests in which half of the doors were openable we obtained the matrix of values shown in Table 11.5.

11.2.3 Collapsing Nodes with Low Refinement Probabilities

Using the refinement probabilities the procedure COLLAPSE-NODES processes the ALPINE-graph. In particular, based on the threshold θ used in the near-DRP condition, Section 11.1, it decides if two nodes should be collapsed into one.

If the refinement probability for both orderings of these nodes is below θ, then no hierarchy in which these nodes are on separate levels will be close to being near-DRP. That is, if $\mathsf{prob}(n_i, n_j) < \theta$ and $\mathsf{prob}(n_j, n_i) < \theta$, as found by the FIND-PROBABILITY routine, then we collapse these two nodes into one $n_{ij} = (n_i \cup n_j)$. This means that when we assign criticalities using a total order produced from this graph the literals in n_i and n_j will be given identical criticalities. Therefore, to ensure we satisfy the constraints imposed by the original partial order we must also collapse all nodes that lie on any path between n_i and n_j into the new node n_{ij}. To collapse the nodes, we modify the graph by substituting all collapsed nodes by the new node n_{ij} and then all in-edges of the collapsed nodes become in-edges of the new node and similarly for the out-edges. The refinement probabilities to and from all the remaining nodes must be recomputed for the new node. To avoid doing this computation, we choose instead to use the average of the original probabilities as estimates for the new ones, so that for every other node n, we let

$$\mathsf{prob}(n, n_{ij}) = \text{Average}\{\mathsf{prob}(n, n_i) \mid n_i \text{ has been collapsed into } n_{ij}\},$$

Table 11.6. Robot box domain hierarchy generated by HIPOINT

Criticality	Predicates
4	Isdoor, Openable
3	Box-Inroom
2	Open
1	Attached
0	Loaded

and similarly for $\text{prob}(n_{ij}, n)$ for nodes n connected via out-edges. The collapsing process continues until no more nodes can be further collapsed.

In robot and box domain the two nodes containing Box-Inroom and **open** will be collapsed to one level whenever the threshold θ is greater than the proportion of openable doors. In this case the refinement probability from Box-Inroom to **open** falls below θ, and the opposite ordering of these nodes is impossible.

11.2.4 Augmented Topological Sort of Abstraction Graph

After collapsing nodes with low refinement probabilities, the procedure AUGMENTED-TOP-SORT computes a total order of the nodes in the resulting graph, using both the partial order relation and refinement probabilities as guides to compute the order. The procedure is a simple modification of a standard topological sort algorithm (see Chapter 4 for a description of the topological sort algorithm). Our augmented version of the algorithm simply orders the collection of nodes that are placed on the queue during any one step. That is, at each stage we add those nodes with no in-edges to the queue, but we add them to the queue so that the sequence of FIND-PROBABILITY estimates between these nodes is ascending. For example, if at some state we are to add the nodes n_1, n_2, and n_3 to the queue we would choose, e.g., the ordering n_3, n_1, n_2 if $\text{prob}(n_3, n_1) \leq \text{prob}(n_1, n_2)$.

To illustrate the procedure, consider again the simple robot example. The graph generated by ALPINE now has associated refinement probabilities, shown in Table 11.5. The augmented topological sort algorithm will place Open right below the Box-Inroom level, resulting in the hierarchy shown in Table 11.6.

11.3 Empirical Results in the Box Domain

In the last section, we have described the HIPOINT algorithm for generating abstraction hierarchies by augmenting the ALPINE algorithm, taking into account the refinement probabilities between abstraction levels. To compare their performance, we have conducted a set of experiments in the box domain, where we placed a time bound of 30 CPU minutes. For the refinement probability estimate computation, a backward-chaining, partial-order planner, ABTWEAK, is used. ABTWEAK is called with a solution length limit of five steps. The features of the domain are that we are able to change the refinement probability at each level individually, by changing the mixture of objects with different properties. For example, we can change the refinement probability between Box-Inroom and open by changing the proportion of openable doors. All systems and domains were implemented in Allegro Common LISP on a Sun4/Sparc Station.

In our realization of the box domain we place two doors between every pair of adjacent rooms. Each door may or may not be openable. The robot's task is to move boxes between the room by either carrying or pulling them.

For domain instances where every door is openable, both ALPINE and HIPOINT generate the same abstraction hierarchy, shown in Table 11.3. Therefore the costs of actual problem-solving using the ALPINE and HIPOINT hierarchies are the same. The only difference is that HIPOINT requires extra time to determine the refinement probabilities. For the domain where every door is openable, HIPOINT takes 27 CPU seconds to generate the hierarchy, while ALPINE took only 0.05 CPU seconds. However, to solve each problem, both systems take 4–5 seconds on average. Thus, as the number of problems grows, HIPOINT's initial cost is amortized away.

We then changed the domain so that not all the doors were openable. In this case not every plan that ignores the status of the doors will be refineable, and HIPOINT will place open higher up in the hierarchy, so that this condition can be tested earlier on, before more resources are allocated to solving the Loaded and Attached preconditions. In our test we set more than 50% of the doors to be openable, so HIPOINT did not collapse any levels. The hierarchy it generated is shown in Table 11.6. HIPOINT takes 75 seconds to generate this hierarchy, after checking 20 randomly generated problems. In contrast, ALPINE is not able to alter its hierarchy in response to a change in the number of openable doors, so it generates the same hierarchy as before.

To compare the qualities of the hierarchies generated by HIPOINT and ALPINE, we first ran ABTWEAK on six problems of varying sizes. ABTWEAK solves each problem by first using the hierarchy generated by HIPOINT, and then by using the one by ALPINE. The CPU time costs for both hierarchies, which do not include the time for generating the hierarchies, are shown in Table 11.7. The table demonstrates that as the planning problems get more complex, HIPOINT is increasingly more efficient than ALPINE. When the initial cost for generating the hierarchies is taken into account, HIPOINT might require a number of problems of small size before it can recover the cost of its more expensive hierarchy generation algorithm. However, for problems of large size, HIPOINT outperforms ALPINE by such a large margin that it can recover its initial cost of hierarchy generation in a single problem, yielding an immediate improvement in net problem solving costs.

Table 11.7. CPU time comparison between HIPOINT and ALPINE in the box domain

Solution Length	HIPOINT(seconds)	ALPINE (seconds)
0	0.08	0.08
3	0.55	0.48
6	4.15	6.55
9	11.17	17.23
12	97.02	419.02
15	229.30	903.25

As a further test we ran 42 test problems of equal length using the hierarchies generated by HIPOINT and ALPINE, respectively. Figure 11.4 compares the accumulated CPU time of both systems over the same 42 problems. The time required by the algorithms to generate their abstraction hierarchies is also included in the values plotted. It is clear from the figure that the HIPOINT hierarchy is soon able to recover from its initial cost, and after 2 problems of this size the net cost was lower than problem solving with the ALPINE hierarchy. In this domain the HIPOINT hierarchy is able to solve problems of this size almost twice as fast as the ALPINE hierarchy, and as the previous table has shown, as the problems become longer so does HIPOINT's factor of improvement.

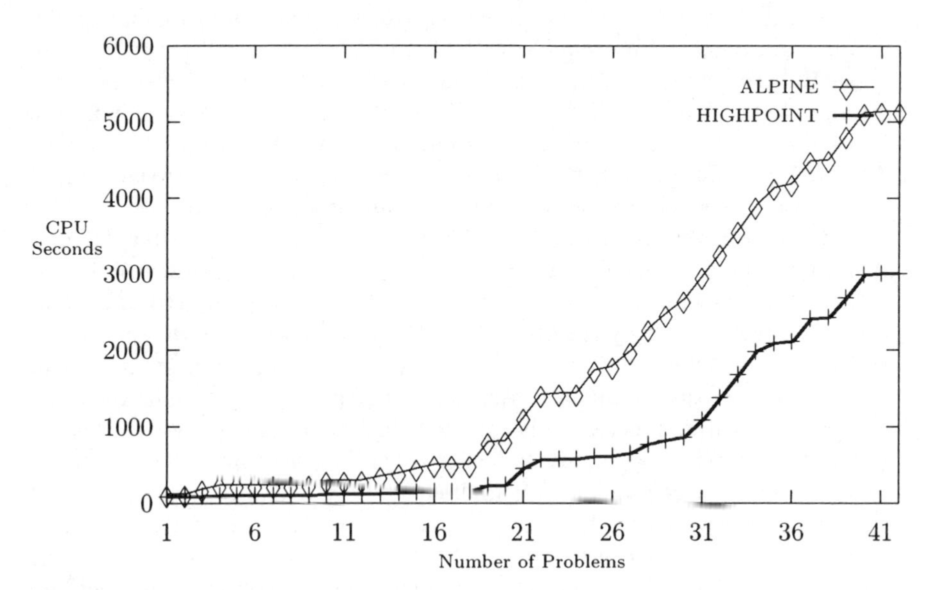

Fig. 11.4. Robot box domain tests

In several other domains we have also shown that HIPOINT is able to offer a significant improvement over ALPINE, which serves to demonstrate the validity of our approach, and the importance of the DRP property. These other results are reported in [13]. To sum up, HIPOINT has the following advantages over ALPINE:

1. As demonstrated by tests in the robot-box domain, when ALPINE generates a partially ordered graph of predicate sets, for hierarchy generation, its selection of a total order does not depend on refinement probabilities. Thus, it is possible that it may generate a hierarchy in which the refinement probabilities are poorly configured, e.g., where they are not increasing as we move down the hierarchy. In contrast, HIPOINT is able to select a more intelligent total order by examining these probabilities.
2. When the refinement probability is low, HIPOINT recognizes the need to collapse two or more levels.
3. Although HIPOINT requires more time to generate its hierarchy, it is clear that over a number of problem solving instances, and for problems with lengthy solutions, this cost is quickly paid off.

11.4 Related Work on Abstraction Generation

In the past, many abstract planning systems have relied on the user to provide an abstraction hierarchy [147, 132, 159]. One of the first systems that semi-automatically generated its own abstraction hierarchies was ABSTRIPS[113].

To build an abstraction hierarchy, in addition to the domain specification, ABSTRIPS also requires a user-defined partial order on the literals in the domain. It then tries to assign criticality values to the literals that are consistent with the given partial order. For each literal l, ABSTRIPS searches for a short plan that achieves l from a state where all literals before l in the partial order hold. If such a plan is found, then l is considered a detail, and is assigned a low criticality value. Otherwise, l will be assigned a high criticality value. This algorithm can be considered as a method for judging the quality of an abstraction hierarchy: a hierarchy is "good" according to ABSTRIPS, if for every low-level literal l, there is a short plan to achieve it from a state where all high-level literals are true. PABLO [31], a successor of ABSTRIPS, can also be viewed in this way.

At a first glance, our syntactic necessary connectivity condition, described in Definition 11.1.2, is similar to ABSTRIPS since they both depend on finding short plans to achieve low level literals. However, there are some significant differences.

First, our condition specifies that all low-level precondition literals of an operator must be *simultaneously* achievable, whereas ABSTRIPS only requires that each literal be achievable individually. However, in domains where interactions often occur, the existence of an individual plan for each literal does not ensure the existence of a plan for the simultaneous achievement of all of the literals.

Second, our necessary connectivity condition specifies that the low-level plan for achieving the preconditions of an operator should not violate any abstract conditions achieved at higher levels. This is in accordance with our notion of monotonic refinement. In contrast, ABSTRIPS does not specify any restriction on the plan that achieves the low level literals. This means that when ABSTRIPS searches for a refinement of an abstract plan it does not restrict itself to searching for monotonic refinements. This can significantly increase the length of the refinement, and can increase the cost of finding it. Also, during refinement the plan to achieve a low level literal might undo work accomplished at the higher level.

In general, although ABSTRIPS attempts to order the literals in such a way as to make its abstract plans easily refineable, it does not take into account all of the factors that affect refineability. Its techniques are mainly heuristic, and were developed without a formal analysis of the problem.

It can also be noted that ABSTRIPS only partially automates the abstraction process; the quality of its hierarchies depends heavily on the user-supplied partial order of the literals.

11.5 Open Problems

HIPOINT can easily be improved, but since this was not the focus of our work we were content with a simple implementation.

The Shortest Solution Assumption

When deriving the exponential speed-up of the DRP in the last chapter, we have assumed that a hierarchical planner using a depth-first search always descends on the shortest solution to the original problem. However, as pointed out by Backstrom and Jonsson [16], this might not always be the case.

To see this, consider an extended TOH domain where we add an additional operator `move-medium-and-large`(?x,?y), for simultaneously moving both the medium disk and the large disk from a peg ?x to ?y. This operator is shown in Table 11.8. With this extension, and with the hierarchy ILMS,

Table 11.8. An additional operator in the extended TOH domain

Preconditions	Effects	Cost
`move-medium-and-large`(?x,?y)		
Onlarge(?x), Onmed(?x) ¬Onsmall(?x), ¬Onsmall(?y) Ispeg(?x), Ispeg(?y)	Onlarge(?y), Onmed(?y) ¬Onlarge(?x), ¬Onmed(?x)	2.0

there are now two interesting abstract plans at Onlarge level. The first plan is the usual moveL(Peg1, Peg3). This abstract plan refines to a 7-step solution plan as obtained in the original TOH domain.

The second abstract solution, move-medium-and-large(Peg1, Peg3), has a cost of 2. Although the cost of this abstract plan is higher than the first one, the plan refines to a less costly concrete-level plan:

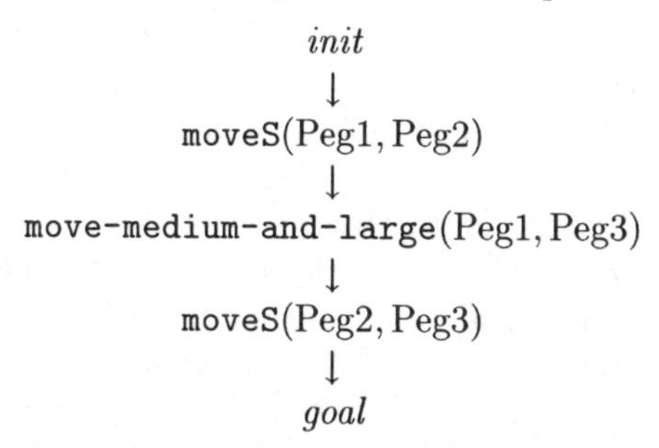

This plan has a cost of 4.

This example reveals one of the shortcomings of HIPOINT and ALPINE. When multiple abstract-level plans exist, a hierarchical planner using an abstraction hierarchy generated by either algorithm would still have to pick one randomly in order to refine the plan to a concrete-level solution. It may be much too late before the planner realizes that it is working on a sub-optimal plan with very high cost. To solve this problem, one could define additional properties to strengthen the constraints placed on a near-DRP hierarchy. Or one could allow a certain amount of look-ahead during refinement of an abstract plan.

Probability Estimate

One way for HIPOINT to be improved is for it to use our analytic forms in Chapter 10 directly to estimate the amount of work required to solve a problem on candidate hierarchies. This would involve a more thorough search through the space of candidate hierarchies; for the searched candidates the analytical forms predicting their behavior could be evaluated using the information gathered by FIND-PROBABILITY. This would give a more accurate evaluation of the hierarchy's worth than the approximations we used.

Another improvement would be to extend HIPOINT to handle individual planning problems rather than to try to construct a single hierarchy for all problems in the domain. Knoblock has demonstrated that problem-specific hierarchies can often display superior performance [80].

Threshold Value

Another source of improvement is to obtain a better understanding of the threshold value θ. In this work, the value is set to 0.5 for near-DRP hierarchies. However, there might be other better values depending on the domain of planning. The appropriate value for θ is likely to depend on the domain features. Finding the right value for θ is therefore an excellent target for machine learning systems.

11.6 Summary

In this chapter we discussed a syntactic condition for testing if a precondition-elimination abstraction hierarchy possesses the DRP. The condition can also be used to estimate refinement probabilities. An algorithm has been provided that automatically constructs an abstraction hierarchy for a given domain. The hierarchies constructed have both the ordered monotonicity property and the near-DRP. Empirical tests have been presented that support our analytical model, and confirm the utility of our algorithm.

11.7 Background

Knoblock's book presents a detailed description of the ALPINE algorithm for constructing an Ordered Monotonic Hierarchy [80]. Smith and Peot [123] provide a critic of the ALPINE algorithm. Lansky and Getoor [87] relate OM hierarchies with problem decomposition using constraints. Bergmann and Wilke [20] argue for an explicit refinement relations between plans at different levels of abstraction. They also present an enlightening review of their PARIS abstraction/case-based system in automated manufacturing domain.

The material of this chapter is largely adapted from an Artificial Intelligence journal article with Bacchus [13]. The ABTWEAK system used in the empirical test is implemented by Woods and Yang (see [149, 160]).

12. Properties of Task Reduction Hierarchies

Along with the precondition-elimination type of abstraction, *hierarchical task-reduction planning*, or HTN planning, is another widely-used hierarchical planning method. Given a set of high-level tasks to be carried out, this method will find ways for reducing each task into subtasks according to a predefined set of task reduction schemata. It then resolves possible conflicts among the subtasks in the plan using a least-commitment strategy. The process is repeated until the tasks in the plan are primitive, and all interactions between the tasks are removed. The advantage is that a hierarchical planner works on a small plan with only an important set of interactions first, before it sets out to handle the rest, which are considered as mere details. This technique has been used in a number of planning systems, including NOAH [114], NONLIN [132], and SIPE [147].

The domain knowledge of a hierarchical planner is organized hierarchically in the form of task-reduction schemata. In addition to a set of primitive tasks, a hierarchical planner also has a set of non-primitive tasks, as well as a set of task-reduction schemata that defines the relationship between the tasks. The efficiency of a hierarchical planner depends on how the task-reduction schemata are defined; if one is not careful in the definition, one may lose the efficiency of hierarchical planning.

This chapter presents the following results.

1. We give a definition for the HTN planning knowledge-base. We specifically define how one operator relates to other operators and causal links via a *reduction* mechanism.
2. We show that with HTN planning, the *upward solution property* is generally not satisfied. This could result in some computational difficulties when using the knowledge-base for planning. In particular, if a planner faces conflicts in a plan that are impossible to resolve by ordering the steps, it cannot abandon the plan right away. Instead it has to try to reduce its steps further to see if interleaving the subtasks introduced at lower levels could resolve the conflicts. This greatly reduces the efficiency of a planner, since it has to search an extra portion of its search space, even if there is strong indication that no solution exists in that space.
3. In response to the above problem, we show that a class of domains exists where the task-reduction schemata can be defined in certain ways, so that

unresolvable conflicts in a plan cannot be resolved in any reduction of the plan. We provide a set of sufficient constraints on schemata definitions to ensure this property.

12.1 Defining Operator Reduction

Task Reductions

A hierarchical planner relies on a set of *operator reduction schemata* Φ. In some sense, these schemata are similar to the operators $\mathcal{O}$ for a single-level, total-order or partial-order planner. A reduction schema $R \in \Phi$ is a function which, when applied to an operator in $\mathcal{O}$, returns a partially-ordered set of operators α_i. R is not necessarily applicable to every operator in $\mathcal{O}$, and an operator can have more than one reduction schema applicable to it. The set of operator reduction schemata applicable to α is denoted by $\text{RSet}(\alpha)$. α is *primitive* if $\text{RSet}(\alpha) = \emptyset$, otherwise it is *non-primitive*. Intuitively, a primitive operator is one which cannot be decomposed further into more detailed steps, while a non-primitive one can.

If R is applicable to α, then $R(\alpha) = \langle A, B, C, D \rangle$, where

- A is a set of operator definitions,
- B defines the binding constraints on the variables,
- C is a set of causal links associated with the specification, and
- D defines a partial ordering among the elements of A.

$R(\alpha)$ is called a *reduction* of α. As an example of operator reduction, consider the household domain. An operator for painting the ceiling and its two reductions are given in Figure 12.1.

For each non-primitive operator α, its reduction schema $R(\alpha)$ defines a *refinement relationship* between α and the set of operators in $R(\alpha)$. Let the latter be $A(R(\alpha))$. The operators in $A(R(\alpha))$ must obey the following restrictions:

Effect Inheritance. Every effect of an operator is asserted by at least one of the operators in its reduction.
That is, $\forall e \in \text{Eff}(\alpha)$, $\exists \alpha_i \in A(R(\alpha))$ such that
- $e \in \text{Eff}(\alpha_i)$, and
- $\forall \beta \in A(R(\alpha))$, if $\neg e \in \text{Eff}(\beta)$ then $\beta \mapsto \alpha_i$. This condition ensures that e is not deleted by some operator in the reduction.

Precondition Inheritance. Every precondition of an operator is also a precondition of at least one of the operators in its reduction, and this precondition is not established by some other operator in the reduction.
That is, $\forall p \in \text{Pre}(\alpha)$, $\exists \alpha_i \in A(R(\alpha))$ such that
- $p \in \text{Pre}(\alpha_i)$, and
- $\forall \beta \in A(R(\alpha))$, if $p \in \text{Eff}(\beta)$ then $\alpha_i \mapsto \beta$. This condition ensures that the precondition p is not internally achieved.

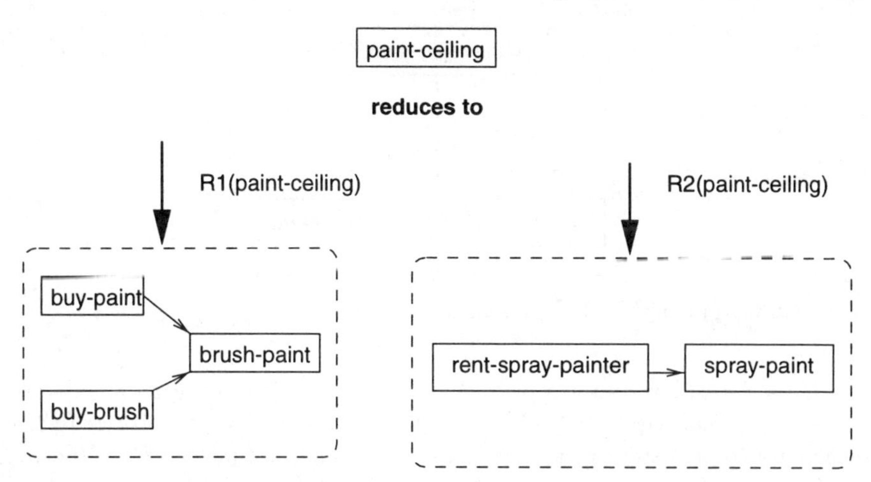

Fig. 12.1. An example operator reduction schema

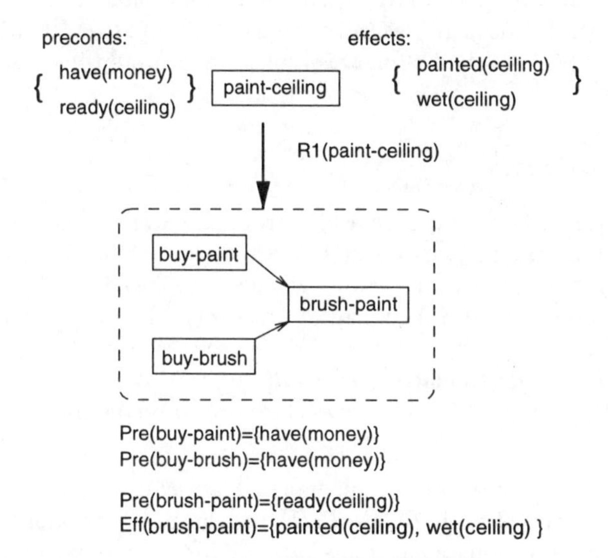

Fig. 12.2. Restrictions on operator reduction schemas

Figure 12.2 shows an example where these two restrictions hold.

In HTN planning we could combine the notions of a goal and a operator into a unified notion of a *task*. In this formalism, a goal is seen as a high-level operator to be achieved, and associated with a goal is a set of skeleton plans that could be used to achieve the goal. These skeleton plans are themselves operator reduction schemata. Figure 12.3 shows an example of goal reduction.

Likewise, in HTN planning, the plan steps are also called *tasks*, so that the reduction process for a goal, operator and plan step becomes unified. In this book, we use the terms "plan steps" and "tasks" interchangeably.

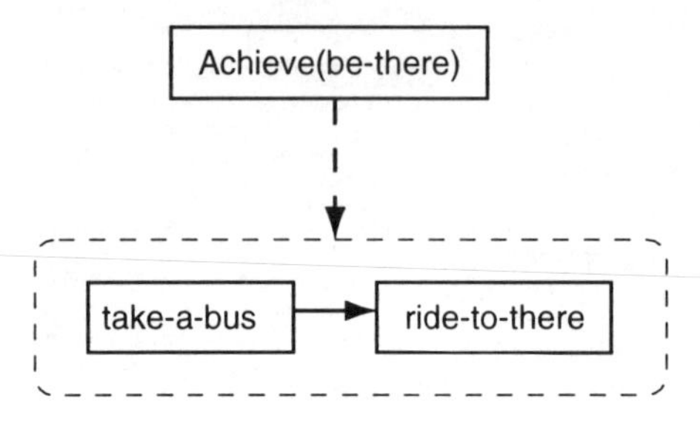

Fig. 12.3. Applying a reduction schema to a goal

Let $\mathbf{s}_i$ be a plan step in a plan Π, resulting from instantiating an operator $\alpha \in \mathcal{O}$ by a substitution θ. We extend the definition of operator reduction schemata to plan steps as follows: for every reduction function R in RSet(α), $R(\mathbf{s}_i) = R(\alpha)\theta$, where $R(\alpha)\theta$ is the result of applying the substitution θ to all operators in $A(R(\alpha))$ (see Chapter 4 for an introduction to substitution), and to all the operator occurrences in the causal links of $R(\alpha)$. If R is applicable to $\mathbf{s}_i$, then $R(\mathbf{s}_i)$ is called a *reduction* of $\mathbf{s}_i$, and a step in $R(\mathbf{s}_i)$ is called a *substep* of $\mathbf{s}_i$.

Plan Reductions

We now consider how to reduce a plan from one level to the next.

Let $\mathbf{s}$ be a step in plan Π, and R a reduction schema applicable to $\mathbf{s}$. Then $R(\Pi)$ is the plan which results when $\mathbf{s}$ is replaced by $R(\mathbf{s})$ in Π. What remains to be specified is how causal links related to $\mathbf{s}$ are transformed in $R(\Pi)$, and how order relations are transposed. We call the former *causal-link-inheritance*, and the latter *order-inheritance.*

With causal-link inheritance, two cases are to be distinguished.

Case 1: $\mathbf{s}$ *is a producer.*
Let $\delta = \langle \mathbf{s} \xrightarrow{p} \mathbf{s}_b \rangle$ be a causal link in Π. After Π is reduced using R, δ is no longer associated with $R(\Pi)$. $R(\Pi)$ will have one or more causal links derived from δ involving a substep $\mathbf{s}_{sub}$ in $A(R(\mathbf{s}))$. In particular, let $\mathbf{s}_{sub}$ be a substep of $\mathbf{s}$ such that

1. p is an effect of $\mathbf{s}_{sub}$, and
2. p is not negated by any other step after $\mathbf{s}_{sub}$ in $R(\mathbf{s})$.

By the definition of R, one or more such $\mathbf{s}_{sub}$ always exists. Then $\langle \mathbf{s}_{sub} \xrightarrow{p} \mathbf{s}_b \rangle$ is a causal link associated with $R(\Pi)$.

Case 2: $\mathbf{s}$ *is a consumer or user.*
Similarly, let $\mathbf{s}$ be a step in a plan Π, and R be a reduction schema applicable to $\mathbf{s}$. Let $R(\Pi)$ be the plan which results when $\mathbf{s}$ is replaced by $R(\mathbf{s})$ in Π. Let $\delta = \langle \mathbf{s}_a \xrightarrow{p} \mathbf{s} \rangle$ be a causal link in Π, and let $\mathbf{s}_{sub}$ a substep of $\mathbf{s}$ such that

1. p is a precondition of $\mathbf{s}_{sub}$, and
2. p is not achieved or negated by any other step before $\mathbf{s}_{sub}$ in $R(\mathbf{s})$.

By the definition of R, one or more such $\mathbf{s}_{sub}$ always exists. Then $\langle \mathbf{s}_a \xrightarrow{p} \mathbf{s}_{sub} \rangle$ is a causal link associated with $R(\Pi)$.

Order-inheritance. Finally, we specify how precedence relations are transposed from a plan to its reductions. Our specification here is adapted from [112].

Let $\mathbf{s}$ be a step in a plan Π, and let $\mathbf{s}_a$ be another step such that $\mathbf{s}_a \mapsto \mathbf{s}$. Suppose that $\mathbf{s}$ is reduced by $R(\mathbf{s})$. Then in plan Π, we require that in $R(\Pi)$, step $\mathbf{s}_a$ is ordered before the *last* substeps of $\mathbf{s}$ in $R(\mathbf{s})$.

Similarly, let $\mathbf{s}_b$ be a step in Π such that $\mathbf{s} \mapsto \mathbf{s}_b$. Then in $R(\Pi)$ it is required that the first substeps in $R(\mathbf{s})$ be before step $\mathbf{s}_b$ in $R(\Pi)$.

Multi-stage Reductions

The above definition for one-step plan reduction can be naturally extended to reduction in more than one step. Let Π be a plan, and $R_1, R_2, \ldots, R_n$ be reduction schemata. Then

$$Q(\Pi) = R_1(R_2(\ldots(R_n(\Pi))\ldots)$$

is a *composite reduction* of Π. Furthermore, if Δ is a set of causal links associated with Π, then $Q(\Delta)$ is the corresponding set of causal links associated with $Q(\Pi)$.

A hierarchical planner operates by iteratively reducing the steps in a plan. Each iteration performs a number of refinements. Algorithm HREFINE (see Table 12.1) depicts how refinement of a plan can be done in this framework. In this algorithm, a procedure HCRITIC takes a plan and applies critics to the plan. A critic can be any plan modification operations, including conflict resolution, plan merging, step reordering, and variable binding.

Table 12.1. The HREFINE algorithm

Algorithm HREFINE(Π, $\mathcal{O}$, Φ)
Input: A plan Π, a set of operators $\mathcal{O}$ and reduction schemata Φ.
External Routines: HCRITIC(Π) applies a set of critics to Π;
Output: Plan reductions.

1. Choose a non-primitive step $\mathbf{s}$ to reduce;
2. SUCC := $\emptyset$;
3. **for each** $R \in$ RSet($\mathbf{s}$) **do**
4. SUCC := SUCC $\cup$ HCRITIC($R(\Pi)$) ;
5. **end for**;
6. **return**(SUCC);

Discussion

Reduction Assumptions. The definition of reduction schemata above is intended to capture the formal aspects of operator or goal expansions in a number of planning systems. In particular, a reduction schema R corresponds to a "soup code" in NOAH [114], an "opschema" or an "actschema" in NONLIN [132], and a "plot" in SIPE [147]. However, several simplifications are made in our formalization of reduction schemata, the first being that "reduction assumptions" are not included explicitly in our definitions. A reduction assumption[1] [29] is a condition on the applicability of a reduction. For example, to clear the top of a block x, a planner may choose a particular reduction that involves two steps: clear the top of block y that is on top of x, then move y to the table. But this reduction is needed only if x is not already clear on its top. In this case, $\mathrm{On}(y, x)$ is a reduction assumption, and an interval is set up that protects the assumption if the reduction is selected. Reduction assumptions are the preconditions of certain substeps in a reduction R that are not achieved by some other substeps in R, and are used by the control component of a planner to restrict the use of certain reductions. The state space generated by all possible reductions that include reduction assumptions is a subset of the state space defined by our formalization. Since the results to be presented below concern the non-existence of solutions in a portion of a state space, it will be easy to verify that all of our subsequent results will hold with the inclusion of reduction assumptions. Thus, we omit them for simplicity.

Planning Level or Abstraction Level. In planning research, the term "hierarchical planning" has been used to describe several different types of planning methods, including subgoaling [102], task reduction [114, 132], and planning at different levels of details [113, 147]. To clarify the concept, Wilkins provides an in-depth discussion of different varieties of planning hierarchies [147], and classifies them into either *planning levels* or *abstraction levels*. A planning level corresponds to the "artifacts of particular planning systems" [147]. For example, a new planning level can be created by expanding each node in a plan according to some pre-defined task-reduction schemata. On the other hand, an abstraction level is distinguished by the "granularity, or the fineness of detail, of the discriminations it makes in the world" [147]. For example, ABSTRIPS [113] is a planner that plans on different levels of abstraction. In this chapter, the term "hierarchy" refers to the planning levels that are obtained by reducing tasks in a plan using a pre-defined set of task reduction schemata.

[1] It is also called a "use-when" condition in NONLIN [132]

12.2 Upward Solution Property

A hierarchical planner functions by repeatedly selecting and reducing the non-primitive plan steps, and applying critics to modify a plan. A critic is a specialized method, either for removing conflicts, or for optimizing the tasks in a plan. Sometimes a critic may find conflicts in a plan that are impossible to resolve. A plan in this situation is said to have *unresolvable conflicts.* Sometimes, backtracking from such a plan is not the best behavior, since although a plan is not correct at the current level of hierarchical planning, it may indeed be possible to make it correct once the plan is reduced to the next lower level.

In Chapter 10 we introduced a property known as the *upward-solution property.* It states that the existence of a concrete-level solution implies the existence of an abstract level solution. If in a hierarchy, the unresolvable conflicts in a plan become resolvable at a lower level, then the upward-solution property is clearly lost. In the following, we first show examples of how the property can be lost in HTN planning.

12.2.1 Losing the Property

We first consider an abstract example. Suppose that a plan contains two non-primitive steps **a** and **b**, such that between them there are no ordering constraints (see Figure 12.4a). Suppose that their effects and preconditions are as follows:

$$\text{Pre}(\mathbf{a}) = \{x\},\ \text{Eff}(\mathbf{a}) = \{u, \neg y\},$$
$$\text{Pre}(\mathbf{b}) = \{y\},\ \text{Eff}(\mathbf{b}) = \{w, \neg x\},$$

where the literals u, w, x, y are all distinct. Then the following conflicts will occur: an effect of **a** deletes a precondition of **b**, and an effect of **b** deletes a precondition of **a**. This kind of conflict is often called a "double cross," meaning that neither ordering of **a** and **b** will work. Unless a planning system can resolve this conflict by inserting additional steps between **a** and **b** to restore the needed preconditions, it will normally announce failure.

However, suppose that the following reductions are available.

1. step **a** can be reduced to the substep $\mathbf{a1} \mapsto \mathbf{a2}$, with $\text{Pre}(\mathbf{a1}) = \text{Pre}(\mathbf{a})$, $\text{Eff}(\mathbf{a2}) = \text{Eff}(\mathbf{a})$, and $x \notin \text{Pre}(\mathbf{a2})$, and
2. step **b** can be reduced to $\mathbf{b1} \mapsto \mathbf{b2}$, with $\text{Pre}(\mathbf{b1}) = \text{Pre}(\mathbf{b})$ and $\text{Eff}(\mathbf{b2}) = \text{Eff}(\mathbf{b})$, and $y \notin \text{Pre}(\mathbf{b2})$.

Then the conflict can be resolved in the reduced plan, by constraining the orderings such that $\mathbf{a1} \mapsto \mathbf{b2}$ and $\mathbf{b1} \mapsto \mathbf{a2}$ (see Figure 12.4b). Clearly, the upward-solution property fails in this hierarchy.

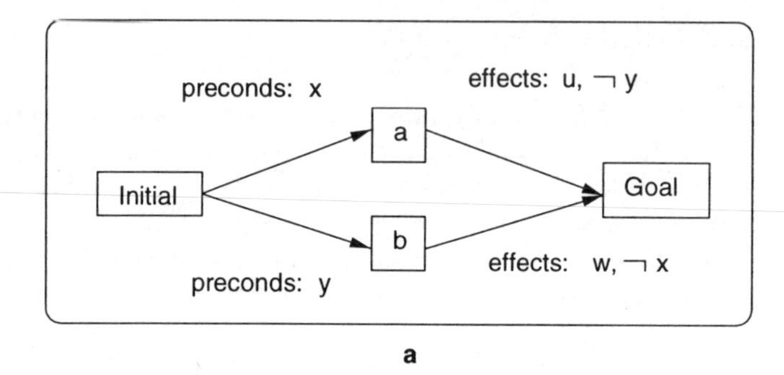

a

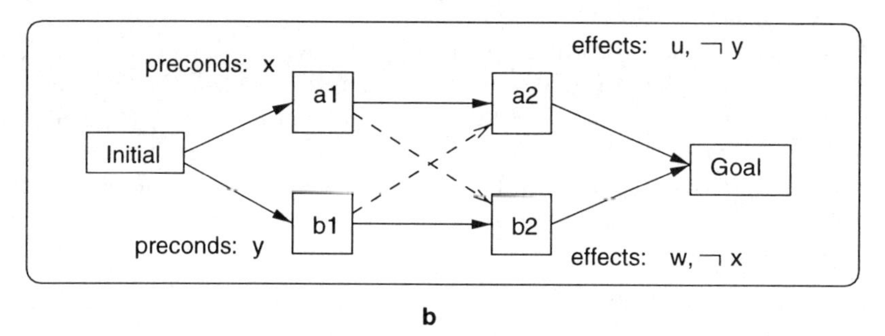

b

Fig. 12.4. (a) A plan with unresolvable conflicts, (b) Resolving the conflict by reducing the plan in (a), by imposing ordering constraints

12.2.2 A Variable-Assignment Example

In the above example, a conflict which appears unresolvable at a higher level is in fact resolvable at a lower level. This happened when both high-level steps are reduced to consecutive substeps.

The problem of losing upward-solution property could occur in other types of reduction as well. Consider the problem of interchanging the values of two variables Figure 12.5. The plan in part (a) contains an unresolvable conflict. If we reduce the `assign-x` step as shown in part (b), however, the conflict could become resolvable by interleaving the `assign-y` step and the two substeps of `assign-x`.

12.2.3 The Chinese Bear Example

The above example shows that when a high-level step splits its preconditions and effects among different substeps, an unresolvable conflict at a higher level could become resolvable. Now we show that this problem could even occur when the effects of a high-level step split between different subtasks.

Consider the following *Chinese bear problem.* According to a Chinese tale, a bear attempted to steal corn from a farmer's field. It decided to transport the corn cobs by placing them under its arms. To begin with, the bear had

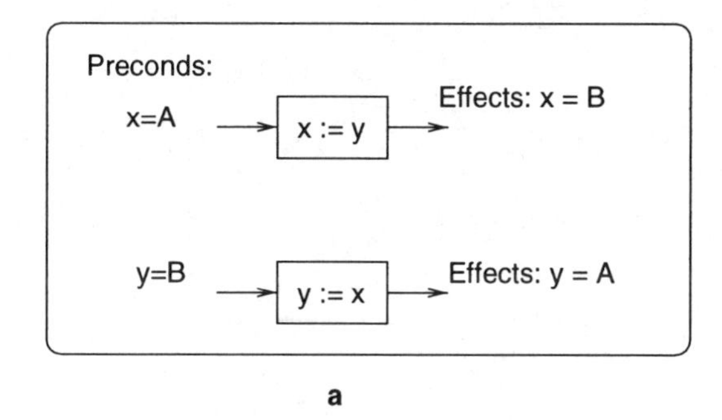

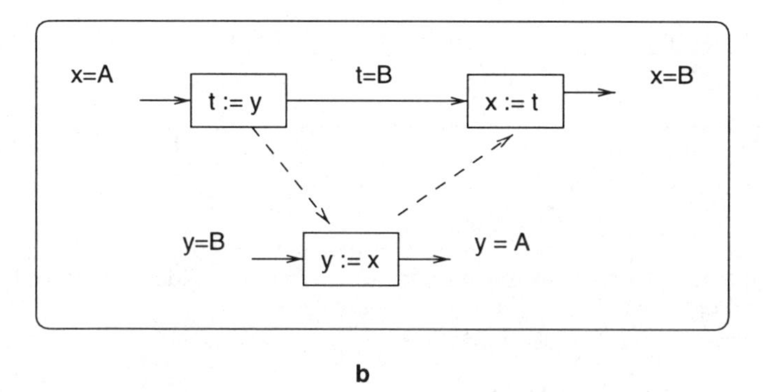

Fig. 12.5. (a) A variable-assignment plan containing unresolvable conflicts, (b) Resolving the conflict by reducing the plan in (a)

two high-level operators, `corn-under-larm` and `corn-under-rarm`, referring to putting the corn under the left arm and the right arm, respectively. A high-level plan consisting of two corresponding steps is shown in Figure 12.6a. This plan contains unresolvable conflicts, since in the process of getting an ear of corn by its *left* hand and placing it under the *right* arm, the bear will lose its "treasure" under the *left* arm, and vice versa.

In the Chinese tale, the bear spent all night trying to resolve this conflict, by repeating the two actions one after another. As day broke, the bear finally had to give up and escape into the forest in haste, empty handed. And, as to the farmer, he woke up with a pleasant surprise that the entire corn field had been neatly harvested!

Our bear could of course do better by simply refining the plan to the next level down, and by interleaving the sub-steps. This is shown in Figure 12.6b. In this plan the bear could indeed put both ears of corn under its arms; it could grab both ears of corn first, one in each hand, then place them under the opposite arms in any order. Using step reduction, the bear could successfully get both corn cobs home.

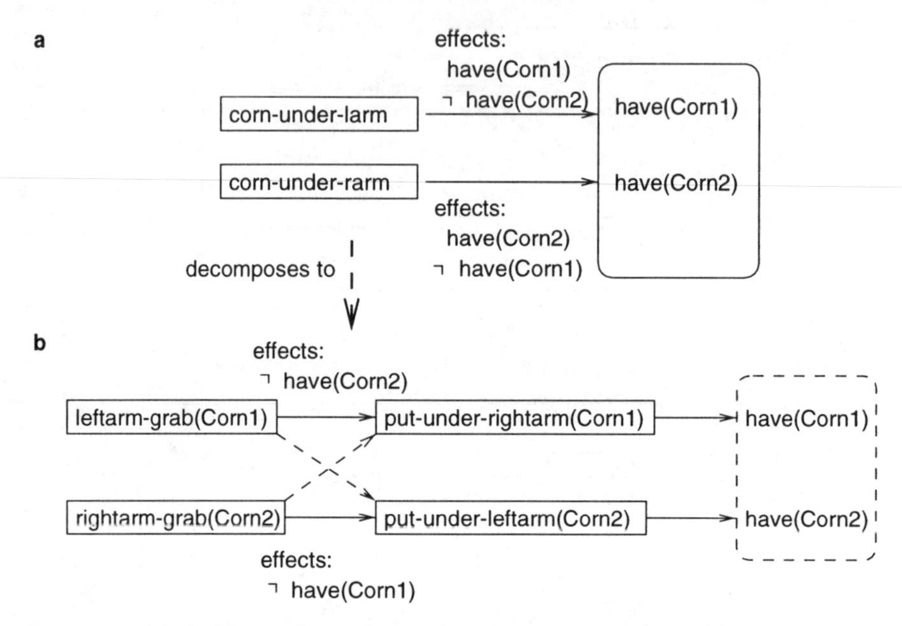

Fig. 12.6. (a) A Chinese bear plan with unresolvable conflicts, (b) Resolving the conflicts by reducing the plan in (a)

12.3 Imposing Restrictions

12.3.1 Motivation

From the search-control point of view, we would like the upward-solution property to hold. If we know that, for a given domain, the operator-reduction schemata satisfy this property, then we could *backtrack* from the current plan whenever the plan contains unresolvable conflicts. However, if the reduction schemata do not satisfy the property, the planner cannot rule out reducing the plan steps further as a means of resolving the conflicts.

One way to guarantee the upward-solution property is to impose restrictions on how the reduction schemata are defined. We would like the restrictions to be *syntactic* in nature, so that checking them against any user-defined reduction schemata could be done automatically. These restrictions may not be satisfied in all domains, but in some cases we could use them to *preprocess* the set of operator reduction schema by redefining the reduction relations. We could also identify the subset of operators which satisfy the restrictions *a priori*.

In the following sections, we will consider one such restriction.

12.3.2 Unresolvable Conflicts

Let Π be a plan. Let C-Links(Π) be the causal links in the plan. We say that Π has unresolvable conflicts if in all total orders of the plan steps, some causal links are violated. That is, in every total order of the plan steps, for some causal link $\langle \mathbf{s}_i \xrightarrow{p} \mathbf{s}_j \rangle$ there exists a plan step $\mathbf{s}_k$, such that

- $\mathbf{s}_k$ is before $\mathbf{s}_j$ and after $\mathbf{s}_i$ in the total order, and
- $\neg p$ is an effect of $\mathbf{s}_k$.

Let Π be a plan with unresolvable conflicts. The upward-solution property is equivalent to the following statement: let Φ be a set of reduction schemata. If $Q(\Pi)$ is any composite reduction of Π using the reduction schemata in Φ, then $Q(\Pi)$ also contains unresolvable conflicts.

If this property is satisfied by a set of reduction schemata Φ, for every plan and every composite reduction of the plan, then we say that Φ is *downward-unresolvable.*

12.3.3 The Unique-Main-Subaction Restriction

We now propose a simple restriction for ensuring the property. The restriction requires that every non-primitive operator has a *unique* main sub-operator, or subaction, that inherits all preconditions and effects of the parent.

Definition 12.3.1 (Unique Main Subaction Restriction). *Let α be an operator and R be a reduction schema applicable to α. The operators in $A(R(\alpha))$ satisfy the following conditions: $\exists \alpha_m \in A(R(\alpha))$ such that*

- *α_m asserts all effects of α, and no other sub-operators of α assert any effect of α.*
  In other words, $\mathrm{Eff}(\alpha) \subseteq \mathrm{Eff}(\alpha_m)$ *and $\forall \alpha_i \in A(R(\alpha))$, such that $\alpha_i \neq \alpha_m$,* $\mathrm{Eff}(\alpha_i) \cap \mathrm{Eff}(\alpha) = \emptyset$.
- *α_m requires all preconditions of α, and none of these conditions is achieved by any other sub-operators of α.*
  In other words, $Pre(\alpha) \subseteq Pre(\alpha_m)$, and $\forall \alpha_i \in A(R(\alpha))$, such that $\alpha_i \neq \alpha_m$, if $\alpha_i \mapsto \alpha_m$ then $\mathrm{Eff}(\alpha_i) \cap Pre(\alpha) = \emptyset$.

This restriction will be referred to as the *Unique-Main-Subaction* Restriction[2]. Intuitively, it requires that, for every step $\mathbf{s}$ in a plan Π, its reduction $R(\mathbf{s})$ contains a "main" substep $\mathbf{s}_m$, which asserts everything $\mathbf{s}$ asserts. In addition, the preconditions of $\mathbf{s}$ are required to "persist" till the beginning of $\mathbf{s}_m$.

A number of planning domains can be represented in ways that satisfy this restriction. For example, consider an operator of fetching an object "object1"

[2] Although for consistency reasons we should call it *unique-main-sub-operator* restriction, we nevertheless choose to follow its original term from [152].

in a room "room1" (based on [147]). The operator "fetch" can be reduced into a sequence of three substeps: getting into room1, getting near object1, and picking up object1, in that order. The third substep, picking up object1, can be considered as the main substep in this reduction. Notice that the condition that object1 is in room1 is a precondition of both the non-primitive operator "fetch" and the main substep. The complete reduction schema is given in Table 12.2.

Let $\mathcal{O}$ be the set of operators of a planning system, and Φ be the set of operator-reduction schemata. We say that Φ satisfies the Unique Main Subaction Restriction if and only if for every reduction schema $R \in \Phi$ and every operator $\alpha \in \mathcal{O}$ to which R is applicable, $R(\alpha)$ satisfies the Unique Main Subaction Restriction. The following theorem can be easily proved by induction on the number of steps inserted into a plan.

Theorem 12.3.1. *Every set of operator-reduction schemata satisfying the Unique Main Subaction Restriction is downward-unresolvable.*

Before discussing how this theorem can be used for checking the downward-unresolvable property, we would like to comment on the applicability of the Unique Main Subaction Restriction to planning domains. Domains that satisfy this restriction have the following characteristics:

1. The goals to be achieved can always be broken down to several less complicated subgoals to solve. For each subgoal, a number of primitive operators are available for achieving it. Each such operator requires several preparation steps before it can be performed, and a number of clean-up steps after it is done. This operator can be considered as the main step in a reduction schema.
2. For each group of operators mentioned above, a hierarchy is built by associating with a non-primitive operator a set of effects which are the purposes of the main step mentioned above, and a set of preconditions which are certain important preconditions of the main step.

A number of domains can be formulated in ways that satisfy the property of downward-unresolvability. Consider again the painting domain. Recall that in this domain there are a room, a ladder, supplies for painting as well as a robot whose goal is to paint portions of the room and/or the ladder. Suppose in addition to the operators and reduction schemata for painting, the robot also knows how to fetch an object using the operator `fetch` in Table 12.2. Represented in this way, every operator-reduction schema satisfies the Unique Main Subaction Restriction. For instance, the `apply-paint-to-ceiling` operator in the reduction of `paint`(Ceiling) is the main sub-operator. Also, the `pickup` step in the reduction of `fetch` is also a main sub-operator. Moreover, these reduction schemata all satisfy the Unique Main Subaction Restriction.

Table 12.2. An operator-reduction schema satisfying the Unique Main Subaction Restriction

preconditions	effects
`fetch`	
Inroom(Object1, Room1)	Holding(Robot1, Object1)
`achieve`(Inroom(Robot1, Room1))	
$\emptyset$	Inroom(Robot1, Room1)
`achieve`(Nextto(Robot1, Object1))	
$\emptyset$	Nextto(Robot1, Object1)
`pickup`	
Inroom(Object1, Room1), Inroom(Robot1, Room1), Nextto(Robot1, Object1)	Holding(Robot1, Object1)

12.4 Preprocessing

One advantage of the Unique Main Subaction Restriction is that it allows for efficient *preprocessing* of a given set of reduction schemata. Specifically, this restriction only limits the way a non-primitive operator relates to its set of sub-operators. Thus the checking process could proceed from an operator to its direct descendents, in an orderly fashion.

To check if $R(\alpha)$ satisfies the Unique Main Subaction Restriction, a test is first made to see if there is a unique main sub-operator in $A(R(\alpha))$ that asserts every effect α asserts and has in its set of preconditions every precondition of α. If so a subsequent check is made on every other sub-operator α_i in $A(R(\alpha))$, to make sure that no other sub-operator of α asserts a precondition or an effect of α.

For each operator and each reduction schema R, verifying the Unique Main Subaction Restriction requires a time complexity of $O(|\mathcal{O}|)$, where $|\mathcal{O}|$ is the number of operators in $\mathcal{O}$. Let k be the maximum number of operator reduction schemata applicable to an operator. Then UMSCHECK has a total worst-case complexity of $O(k \times |\mathcal{O}|^2)$.

If the UMSCHECK returns TRUE, then the set Φ of reduction schemata satisfies the downward-unresolvability restriction. In that case, whenever a hierarchical planner detects unresolvable conflicts in a plan, it does not have to consider the reduction of the plan as a means of resolving the conflicts.

12.5 Modifying Operator-Reduction Schemata

Even if UMSCHECK returns FALSE for Φ, in some cases it might still be possible to modify the schemata, so that after the modification, Φ satisfies the restriction.

Table 12.3. The UMSCHECK algorithm

Algorithm UMSCHECK($\mathcal{O}$, Φ)
Input: A set of operators $\mathcal{O}$ and reduction schemata Φ.
Output: TRUE iff the unique-main-subaction restriction is satisfied.

```
1.   B := O;
2.   while B ≠ ∅ do
3.     α := Remove-element(B);
4.     if ∀R ∈ RSet(α),
5.     R(α) satisfies the Unique-Main-Subaction Restriction then
6.       Mark α;
7.     end if
8.     B := B − {α};
9.   end while
10.  if all operators in O are marked then
11.    return(TRUE);
12.  else
13.    return(FALSE);
14.  end if
```

We illustrate the modification process via Figure 12.7. In the figure, α is an operator with a precondition p and an effect q. Suppose $R(\alpha)$ consists of two sub-operators $\alpha 1$ followed by $\alpha 2$, such that the precondition of $\alpha 1$ is p, and the effect of $\alpha 2$ is q (see Figure 12.7a). Clearly, the Unique Main Subaction Restriction is violated. One way to modify this reduction schema is to remove the precondition p from Pre(α). As a result, the modified reduction R' satisfies Unique-Main-Subaction Restriction. This new reduction schema is shown in Figure 12.7b.

The above example shows how to modify a reduction schema by removing the preconditions of an operator. Other methods for schemata-modification can also be designed. For instance, let α be an operator with two effects q and u such that a sub-operator $\alpha 1$ of α has an effect q while another sub-operator $\alpha 2$ has an effect u (see Figure 12.8a). If u is considered as a side-effect of α, and is considered to be less important than q, then removing u from the the effects of α will make the reduction schema satisfy the Unique Main Subaction Restriction. This is shown in Figure 12.8b.

Unfortunately, the above examples do not suggest any general procedure for modifying a given set of operator-reduction schemata. This is because changing the representation of the operators in $\mathcal{O}$ may make them less expressive than before. This can occur when, for example, some effects of an operator are removed from its effects set. In extreme cases, changing the representation of an operator may cause planning to be less efficient than before. For example, removing the preconditions from an operator may delay the detection of deleted-condition conflicts with the removed conditions, and as a

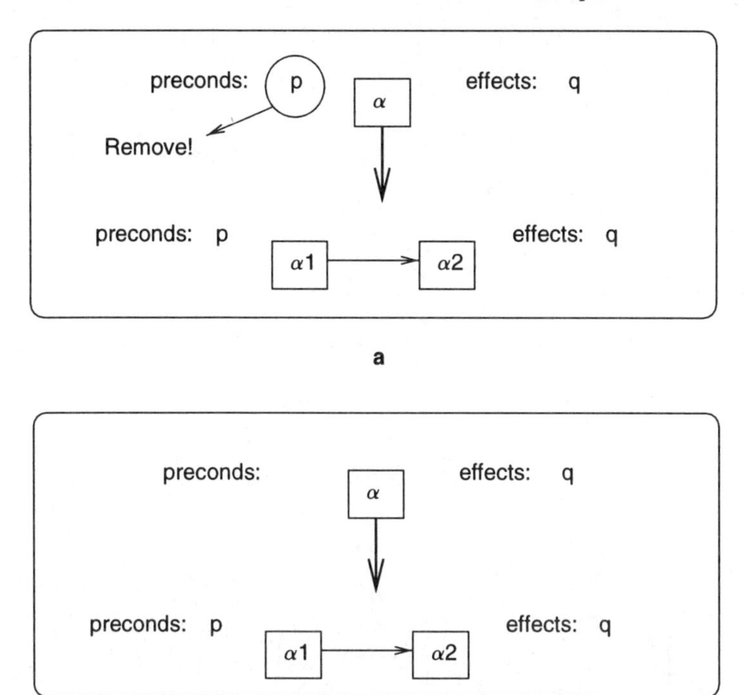

Fig. 12.7. Modifying a reduction schema (a) by removing some preconditions of an operator (b)

result, more effort in reasoning may be needed to fix those conflicts after the modification. On the other hand, there may exist classes of planning problems for which a particular kind of schemata modification method exists, and finding out the characteristics of these domains would be a very interesting exercise.

12.6 Open Problems

More Expressive Languages

So far we have considered hierarchical planning with a particular kind of operator representation, in which the preconditions and effects of an operator are literals. This restricts the techniques developed to be only useful for a subset of the existing hierarchical planning systems. For example, SIPE [147] allows for conditions which can be context-dependent, quantified, and disjunctive. In addition, some effects can be deduced from a set of external axioms. With operator representations as expressive as these, some of our results no longer apply.

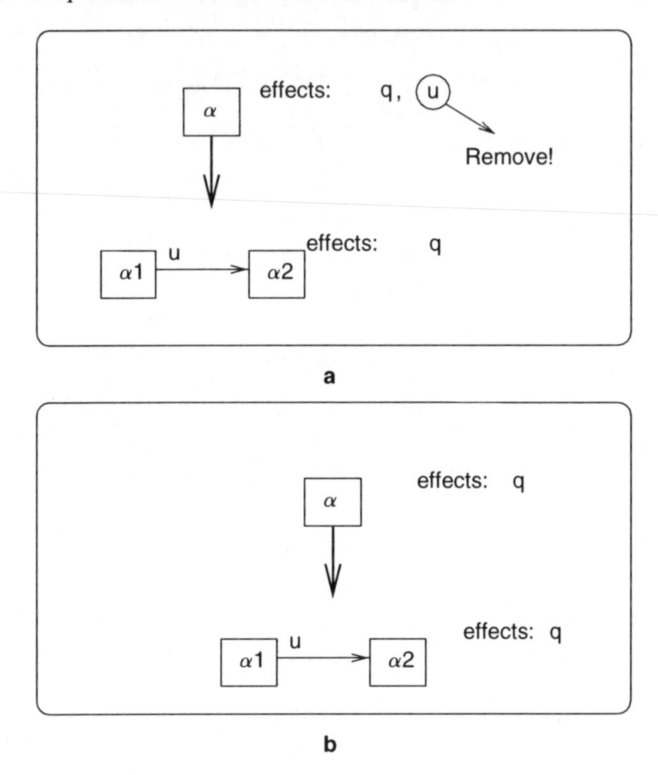

Fig. 12.8. Modifying a reduction schema (a) by removing effects of operator α (b)

For example, consider operators with preconditions in the form of disjunctions of conditions. Then the effects of a group of operators can collectively deny or assert the preconditions of other operators. With these representations, the detection of conflicts during planning is a more complicated problem. It is believed that the idea of extracting information about conflict *a priori* can be used to help ease this computational burden, in the same way this chapter has shown. For example, information about which operators are more likely to interact may be obtained through preprocessing and used to provide a better control strategy for the detection and handling of conflicts.

Heuristics for Planning

Preprocessing can also provide a planner with good heuristics for choosing among alternative operator-reduction schemata. To reduce a non-primitive operator, there is usually more than one available schema, and current hierarchical planning systems select the alternatives in a depth-first manner. Depending on which reductions are chosen, different conflicts may be introduced, and this may affect the planning efficiency.

One heuristic which can be used in choosing a reduction is to minimize the number of conflicts introduced. Preprocessing can be helpful for predicting which reductions are likely to yield a set of conflicts that is small—and just as importantly, a set of conflicts which are easy to resolve.

12.7 Summary

In this chapter we have outlined a formalization of hierarchical planning under task reduction. Based on the formalism we have also developed a syntactic restriction on the relationship between a non-primitive operator and its set of sub-operators. When satisfied, the restriction enables hierarchical planning systems to recognize a dead-end state earlier. Conflicts in a plan can then be concluded as unresolvable at every level of reduction below, given that they are unresolvable at the current level. Because of the syntactic nature of the restrictions, it is possible to preprocess a given set of reduction schemata.

12.8 Background

The basic structure of hierarchical-task network planning was determined by two influential systems, Sacerdoti's NOAH [114] and Tate's NONLIN [132]. Wilkins gave a comprehensive overview in [147].

There is an increasing amount of interest in HTN planning, particularly due to its ability to effectively encode much knowledge about both problem decomposition and hierarchical abstraction. On-going investigations include papers appearing in AIPS94 [161] and AAAI94 ([18], [43], and [133]).

The material of this chapter is adapted from a Computational Intelligence journal article [152], which in turn was an adaptation of part of Yang's PhD thesis [150]. Nau and Hendler provided much valuable advice to the latter.

13. Effect Abstraction

In this chapter we are interested in identifying the *primary effects* of planning operators for the purpose of improving planning efficiency and maintaining solution quality. In many realistic domains, a planning operator might have a large number of effects; only a few of which are relevant to the problem at hand. The idea of planning with primary effects is to specify certain effects of planning operators as *primary*, and to choose operators for achieving goals only in terms of the operators' primary effects. Identifying primary effects among operators is analogous to building a two-level abstraction hierarchy on the effects of operators. The primary effects are on the top level. During plan synthesis, only the top-level effects are considered for achieving subgoals. Consideration of all effects is done only when plan correctness is verified.

This chapter presents a formal model of planning and learning with primary effects.

13.1 A Motivating Example

The intuition behind primary effects is quite natural. Consider the household domain for example. A painting operator might have many different effects, as shown in Figure 13.1, ranging from Painted(Wall) to Exercised(Agent). Among those effects, we would normally consider Painted(Wall) as a primary effect of painting. The other effects, such as Exercised(Agent), are considered *secondary*. Put it another way, if an agent only wants to exercise, he or she could choose a much cheaper and more pleasant way of doing it, by jogging,

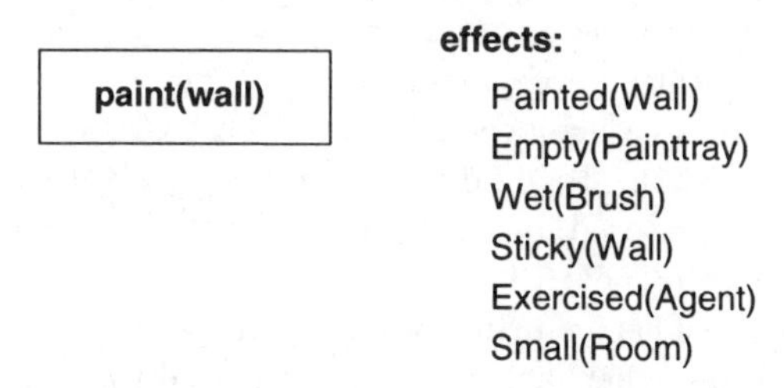

Fig. 13.1. A painting operator

riding a bike, etc. Painting is invoked only when Painted(Wall) is posed as a goal to achieve. As another example, the primary effect of breaking a hole in an office wall is to create a door or a window. If we merely wanted to go to another office, we could do it more easily by simply walking to that office through a doorway.

Primary effects have been used in planning in a fairly informal way in several planning systems.

- STRIPS [46] utilizes primary effects as a means to control the quality of solution plans.
- SIPE [147] distinguishes between the "main effects" and the "side effects" of operators and uses this distinction to simplify conflict resolution in planning.
- PRODIGY [25, 80] allows the user to specify primary effects of operators, which are then used as an effective method to control search.
- ABTWEAK [160] also allows the user to specify primary effects, the use of which significantly improves the efficiency of planning.

In all these systems human designers are relied upon to identify the primary effects, and no effort has been made to build a well-founded guidance for the selection of those effects.

In this chapter, we address several important issues involved in understanding primary effects. First, we present a representative algorithm for planning with primary effects. Then we analyze the effectiveness of using primary effects in improving search efficiency and solution quality. Finally, we present a learning algorithm for automatically identifying the primary effects of operators and analyze its effectiveness.

13.2 Primary-Effect Restricted Planning

Given that primary effects of planning operators have been chosen, it is in fact very easy to modify a partial-order planner to obtain one that uses only primary effects to achieve subgoals. To see this, let POPLAN be a backward-chaining, partial-order planner that does not distinguish between primary and secondary effects, and let PRIMPOP be its primary-effect counterpart. We will call PRIMPOP a *primary-effect restricted planner*, and the original POPLAN planner an unrestricted planner.

In most implementations of partial-order planners, a key component is how to select operators to achieve subgoals. These steps are shown characteristically in Table 13.1a. To modify the planning algorithm so that only primary effects are considered for achieving subgoals, we need to modify only a few lines, as shown in Table 13.1b.

A major purpose of using primary effects in planning is to restrict the number of possibilities when achieving a subgoal. In terms of search, this corresponds to *reducing the branching factor* of the search tree.

Table 13.1. Comparing unrestricted and primary-effect restricted planners. The differences between the two planners are highlighted by boldface

a	b
Algorithm POPLAN(Π, $\mathcal{O}$, ...);	**Algorithm** PRIMPOP(Π, $\mathcal{O}$, ...);
Input:	**Input:**
An initial plan Π,	An initial plan Π,
a set of planning operators $\mathcal{O}$	a set of planning operators $\mathcal{O}$,
etc.	**with chosen primary effects**, etc.
Output:	**Output:**
A correct plan if there is one;	A correct plan if there is one;
Fail otherwise.	Fail otherwise.
...	...
...	...
/* *achieve new subgoals* */	/* *achieve new subgoals* */
choose a subgoal $(p, \mathbf{s})$ from Π,	choose a subgoal $(p, \mathbf{s})$ from Π,
where p is an open precondition of $\mathbf{s}$;	where p is an open precondition of $\mathbf{s}$;
a) find all operators α in $\mathcal{O}$	**a)** find all operators α in $\mathcal{O}$
such that an **effect** of	such that a **primary effect** of
α unifies with p;	α unifies with p;
b) find all existing steps $\mathbf{s}_i$ in Π	**b)** find all existing steps $\mathbf{s}_i$ in Π
such that an **effect** of	such that a **primary effect** of
$\mathbf{s}_i$ unifies with p, and	$\mathbf{s}_i$ unifies with p, and
that $\mathbf{s}_i$ can be ordered before $\mathbf{s}$;	that $\mathbf{s}_i$ can be ordered before $\mathbf{s}$;
for each such operator α	**for each** such operator α
or step $\mathbf{s}_i$ **do**	or step $\mathbf{s}_i$ **do**
update plan Π, resulting in	update plan Π, resulting in
a successor plan;	a successor plan;
...	...
...	...

As an example, consider an office domain with an agent and several objects that can be moved from room to room. There are several operators in this domain; a few of them are depicted in Table 13.2. An agent can move around offices by simply walking from one room to another. However, there are also other ways of moving the agent around — by carrying a book, or by pushing a trolley. In order to just go to another room, the simplest way is to walk without carrying a book or using a trolley. From this intuition, an assignment of primary effects as shown by *'s in Table 13.2 seems natural.

An example search tree for achieving the goal of Inroom(Agent, Room2) is shown in Figure 13.2. Part (a) shows a search tree of an unrestricted planner, where all effects of all operators are considered for achieving a subgoal. Part (b) shows a tree corresponding to a primary-effect restricted planner. This tree is considerably narrower because the nodes at each level of the tree are a subset of the nodes at a corresponding level of the tree in part (a).

Table 13.2. Operator definition, with selected primary effects, for the office domain example

Precondition	Effect	Cost
`walk`(Agent, Room1, Room2)		
Handempty, Inroom(Agent, Room1), Open(Door)	* Inroom(Agent, Room2), ¬Inroom(Agent, Room1)	1.0
`carry`(Agent, Book, Room1, Room2)		
Holding(Agent, Book), Inroom(Book, Room1), Inroom(Agent, Room1) Open(Door)	* Inroom(Book, Room2), ¬Inroom(Book, Room1), Inroom(Agent, Room2), ¬Inroom(Agent, Room1)	5.0
`push-trolley`(Agent, Trolley, Book, Room1, Room2)		
Ontrolley(Book), Inroom(Book, Room1), Inroom(Agent, Room1), Inroom(Trolley, Room1), Open(Door)	Inroom(Book, Room2), ¬Inroom(Book, Room1) Inroom(Agent, Room2), ¬Inroom(Agent, Room1) ∗Inroom(Trolley, Room2), ∗¬Inroom(Trolley, Room1)	10.0

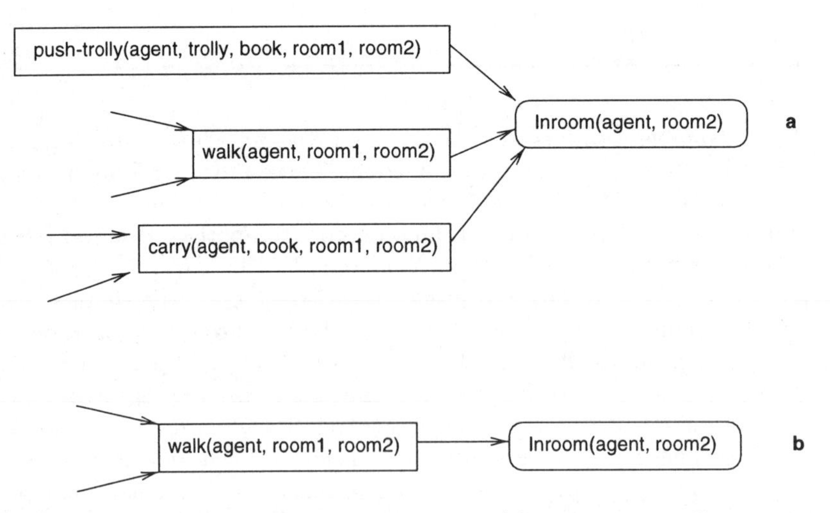

Fig. 13.2. Search trees of unrestricted planning (a) and primary-effect restricted planning (b)

This example shows a general fact about using primary effects: primary-effect restricted planners usually have smaller branching factors. This implies that, if we somehow limit the depth of the tree, a primary-effect restricted planner could expand a smaller search tree, and is thus more efficient than an unrestricted planner.

The same example shows another point about using primary effects. Suppose that one uses a *depth-first search* control strategy to operate an unrestricted planner. The planning process then corresponds to searching the tree in Figure 13.2a in a top-to-bottom order. If the first branch involving a trolley actually leads to a solution, it is conceivable that this solution is much more costly than the one which involves walking only. On the other hand, if a primary-effect restricted planner is used, a tree similar to the one in Figure 13.2b is searched. The first solution, again in a top-to-bottom order, is simply the optimal solution in this case. Thus, in this case, using primary effects also improves the solution quality. This is the main motivation behind STRIPS in its use of primary effects.

In general, proper uses of primary effects in planning can lead to higher planning efficiency and improved solution quality. This conclusion, however, corresponds to a best-case scenario; it assumes that the primary effects are chosen appropriately. In the next section, we show that an arbitrary selection of primary effects could lead to an erosion of planning efficiency as well as solution quality.

13.3 Incompleteness and Sub-optimal Solutions

Using primary effects in planning corresponds to cutting down the number of possible plans to achieve a given goal. This is because by restricting a planner's attention only towards the primary effects, all operator sequences that achieve a goal based on the *side effects* of the operators will not be part of the search tree of the planner. The implication is that if primary effects are chosen poorly, a primary-effect restricted planner may not be able to find a good solution to a planning problem and, in some cases, may not find a solution at all.

As an example, in the planning problem shown in Figure 13.3, suppose that each square represents a plan step, and every path in the directed graph represents a solution to the planning problem. The crosses represent those branches that are considered by an unrestricted planner, but are not considered by a primary-effect restricted planner. If the number of steps (that is, squares) in a path, from start to finish, represents the cost of a plan, then the particular selection of primary effects which gives rise to the remaining path in the graph leaves a longer, sub-optimal solution plan to the planning problem. In this example, we see clearly that an arbitrary selection of primary effects might lead to sub-optimality.

In extreme cases, a poor selection of primary effects could also lead to *incompleteness* of planning. In the household example, if painting is the only

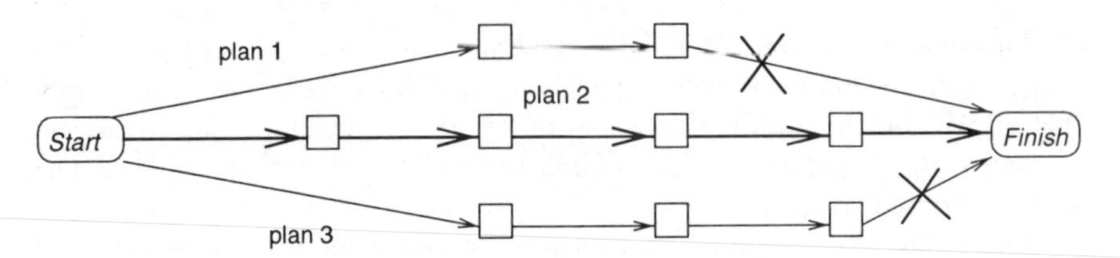

Fig. 13.3. Demonstrating problems of primary-effect restricted planning

means for the agent to do exercise, and if the effect Exercised(Agent) is not chosen as a primary effect of operator `paint`, then the agent will have no solution plan for doing exercise.

We summarize these two problems as follows:

Incompleteness. Inability to find a solution to a solvable planning problem.
Sub-Optimality. Even if a solution plan can be found, a primary-effect restricted planner might only find a poor solution to a planning problem.

In the next section, we will describe an approach to solving these two problems. Our solution will be to use a learning algorithm, which will perform a series of training exercises to obtain a *good* bias on its primary effect selection.

13.4 An Inductive Learning Algorithm

Starting with a set of operators for which no primary effects have been selected, we would like to design a learning algorithm for constructing a selection of primary effects based on a set of training examples. Each example is a planning problem typically seen by the planner, and the set of examples could be built based on the past experiences or on randomly generated training sets. After learning, we would like the learned primary effects to have good quality. Specifically, we would like to approach a selection of primary effects that ensures reduced planning time and near-optimality in solution quality, on average.

The intuition behind the learning algorithm is as follows. For each planning problem, consisting of a start-step/finish-step pair, we run an optimal, unrestricted planner and a primary-effect restricted planner side by side. If and when both planners are finished, we then compare their solution quality. Let Π_o be the optimal plan returned by the unrestricted planner, and Π_p a plan returned by the primary-effect restricted planner. Then we check whether the following conditions are satisfied:

$$\text{Cost}(\Pi_p) \leq C * \text{Cost}(\Pi_o) \tag{13.1}$$

In the above equation, the factor C is a user-defined *cost-increase value*, representing how much the user is willing to tolerate a degradation in solution quality; a solution Π_p is near-optimal if its cost is a C-factor to the optimal cost. For instance, when $C = 3$ it means that the user is unwilling to have a primary-effect restricted planner return solutions that are more than three times as costly as the optimal solutions.

If (13.1) is satisfied, then the learning algorithm will go on to test the next problem. Otherwise, the learning algorithm will convert additional effects of operators to primary. The rationale for this operation can be informally explained as follows. Suppose that condition (13.1) is violated. This could be due to a higher cost in the solution plan of the primary-effect restricted planner. Now recall that the branching factor of a primary-effect restricted planner is generally smaller than that of its unrestricted counterpart, whereas the search depth of the former is generally larger than that of the latter. Converting effects into new primary effects will have the impact of increasing the search space size for a primary-effect restricted planner. The solution quality in a larger space will be no worse than that in the space before the conversion. Therefore, condition (13.1) will be satisfied by more samples the next time around.

We now describe the algorithm formally. The LEARNPRIM algorithm, shown in Table 13.3, takes a randomly generated problem in each iteration. It then computes a plan using an optimal unrestricted planner OPTIMALPOP and then using a primary-effect restricted planner PRIMPOP. If both planners terminate within a reasonable time bounds, the algorithm verifies condition (13.1) using the accumulated time and cost information. In the event the condition fails, LEARNPRIM calls a subroutine UPDATEPRIM, which turns all effects in the causal links of Π_o to primary effects, for operators that correspond to the steps producing these effects. The LEARNPRIM algorithm stops learning when a certain termination condition TERMINATE holds. In the next section, we derive this termination condition.

We illustrate the operation of this algorithm through the robot example shown in Table 13.2. Suppose that initially no primary effect is chosen. Suppose also that the user has provided a set of two training examples:

$$\langle\ \mathit{init},\ G_1\ \rangle,\ \langle\ \mathit{init},\ G_2\ \rangle$$

where *init* is an initial state in which both the agent and a book are in Room1, $G_1 = \{\text{Inroom(Agent, Room2)}\}$ and $G_2 = \{\text{Inroom(Book, Room2)}\}$. Also, let $C = 2$.

For the first problem $\langle\ \mathit{init},\ G_1\ \rangle$, an optimal solution found by POPLAN is

$$\Pi_1 = (\mathtt{s}_{init} \mapsto \mathtt{walk}(\text{Agent, Room1, Room2}) \mapsto \mathtt{s}_{finish})$$

Since at this point no primary effects have been chosen, PRIMPOP will announce failure. In this case, Π_p has a cost of infinity, and condition (13.1) fails. UPDATEPRIM will then examine the causal links of Π_1, and find out

Tablc 13.3. The LEARNPRIM algorithm

Algorithm LEARNPRIM($\mathcal{O}$)
Input: A set of planning operators $\mathcal{O}$.
External Function:
NEXTPROB() returns an initial plan consisting of a randomly generated start-finish step pair.
TERMINATE() is a boolean function. See the next section for its specification.
Output: $\mathcal{O}$ with selected primary effects.

1. **loop**
2. Π_{init} := NEXTPROB();
3. Π_o := OPTIMALPOP(Π_{init});
4. Π_p := PRIMPOP(Π_{init});
5. **if** Condition 13.1 is not TRUE, **then**
6. UPDATEPRIM($\Pi_o, \mathcal{O}$)
7. **end if;**
8. **until** TERMINATE()=TRUE;
9. **return**($\mathcal{O}$);

Subroutine UPDATEPRIM(Π_o, $\mathcal{O}$)

1. **for each** $(\langle \mathbf{s}_i \xrightarrow{p} \mathbf{s}_j \rangle) \in$ C-Links(Π_o) **do**
2. let α be the operator in $\mathcal{O}$ corresponding to $\mathbf{s}_i$;
3. make p a primary effect of α;
4. **end for**

---

that the plan step `walk` produces the precondition Inroom(Agent,Room2) for $\mathbf{s}_{finish}$. This effect of `walk` will be a new primary effect.

Continuing with the second planning problem in the same manner, the effect Inroom(Book, Room2) will become a primary effect of operator `carry`. Suppose that these two problems are the only examples supplied by the user, then at some point (see the next section), the learning algorithm would terminate. The end result is shown in Table 13.2.

This example shows another interesting feature of the learning algorithm. By learning primary effects, certain *redundant operators* could be identified and removed from the operator set $\mathcal{O}$. In the robot example the redundant operator is `push-trolley`. Removing these operators could further improve the efficiency of planning.

13.5 How Many Training Examples

At this point, it has not been specified how the examples are generated, and how many examples need be used to train the planner. We answer these questions below.

13.5.1 Informal Description

The LEARNPRIM algorithm depends on a routine NEXTPROB() for generating the next problem instance. A problem instance is an initial state and goal pair. While these problems are generated randomly, an important requirement is that the *probability distribution* of the examples used in the learning phase must be the *same* as the distribution of the problems to be solved later by the primary-effect restricted planner, using the selected primary effects. To obtain the distribution, we can use a pool of existing plans that are solved by an unrestricted planner in the past. Each of these plans is associated with a cost, an initial state, and a goal. The relative frequency of each problem instance could then be used to approximate the probability distribution.

Another method for the generation of training examples is to rely on a *query generator.* The query generator is a potential user of the primary-effect restricted planner, and will pose planning problems to be solved at a given, fixed-probability distribution. As an example, a frequent database user could be considered as a query generator; so is a typical client of an information provider.

Suppose that we fix a method for implementing NEXTPROB(). We now consider the second question — how many examples are needed. Informally, each example *(Initial, Goal)* is a point in a space of problem instances (see Figure 13.4). After an iteration of LEARNPRIM, the example *(Initial, Goal)* is considered learned. That is, the next time we run PRIMPOP on the same example, condition (13.1) will be satisfied. In addition, due to the newly selected primary effects, a few other problem instances that are near *(Initial, Goal)* could potentially be covered as well. Thus, each training example has a certain *coverage* of learned examples in the space of all problem instances. This is shown as a circle around the *(Initial, Goal)* in Figure 13.4. In answering how many examples are needed, it is hoped that, with enough examples, most of the area where a problem occurs with *high probability* will eventually be covered.

In fact, many of the intuitions about the problem coverage have been addressed in a formal learning model known as *PAC Learning.* In the following, we briefly review the main points in this computational model, and then apply the model to our problem at hand.

13.5.2 A Brief Introduction to PAC Learning

"PAC learning" is a shorthand for *probably approximately correct* learning. It is an inductive learning method for hypothesis selection. Let X be a set of

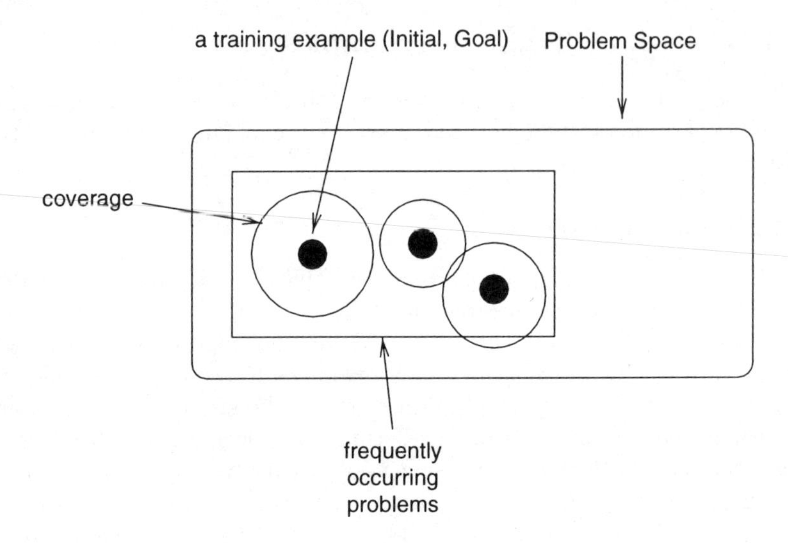

Fig. 13.4. Informal description of the learning process

all possible examples. X is known as the instance space. A concept over X is a subset of X, representing some features satisfying a certain condition. A concept class H over X is the set of all concepts on X; this is known as the hypothesis space on X. Each element h of H is a hypothesis. A hypothesis is similar to a predicate: for each $x \in X$, $h(x) = \text{TRUE}$ if and only if x is a member of the subset of X defined by h.

Let D be an possibly unknown probability distribution on the instance space X. Every element x in X has a probability of occurrence, $\mathbf{Pr}_D(x)$. For a subset S of X, the probability for S to occur is

$$\sum_{x \in S} \mathbf{Pr}_D(x)$$

Let t be a target concept to be learned. The elements x of X are classified by t into positive examples, for which $t(x) = \text{TRUE}$, and negative examples for which $t(x) = \text{FALSE}$.

It is sometimes impossible to learn the exact target concept t. So we might settle for an *approximate concept* or hypothesis h, such that the *error* in h is small. With the above definitions, the error in a hypothesis could be defined formally:

$$\mathbf{error}(h) = \mathbf{Pr}_D\{x | h(x) \neq t(x)\}$$

A hypothesis is *approximately correct* if $\mathbf{error}(h) < \epsilon$ for some user-defined small and positive real-value ϵ.

The goal of PAC learning is to learn an approximate concept by sampling the elements of X according to the distribution D, for a finite number of times. Each training example x is supplied with a correct answer, $t(x)$. For a given sequence of training examples, a hypothesis h is called *consistent* if for all training examples x in the sequence, $h(x) = t(x)$.

The training samples are taken randomly, so there is no absolute guarantee that the examples seen by a learning algorithm will be typical enough to be useful. In this case, we settle for a *probably approximate correct* concept, in the following sense. Let δ be small and positive real value. We wish to guarantee that, after enough samples are used in training, with probability no less than $1 - \delta$, the concept h we learn is approximately correct.

Let m be the number of examples used to learn a probably approximately correct concept h. We wish m to be finite and reasonably bounded. There is a fairly simple but very powerful result in PAC learning, which states that if the hypothesis space H is bounded, then for any sufficiently large m, any consistent concept h is a PAC concept.

We derive this result. Suppose that h' is a bad hypothesis; $\mathbf{error}(h') > \epsilon$. Suppose that h' is consistent with all m examples. The probability that h' agrees with any individual example is less than $(1 - \epsilon)$, and the probability that h' will be consistent with all m examples is less than $(1 - \epsilon)^m$. The probability for a bad hypothesis to be consistent is $|H|(1 - \epsilon)^m$. For this to be less than δ,

$$|H|(1 - \epsilon)^m < \delta$$

This formula gives rise to an upper bound on m:

$$m \geq \frac{1}{\epsilon}(\ln \frac{1}{\delta} + \ln |H|)$$

Thus, a hypothesis h is PAC if it is consistent with m examples, where m is given as above.

13.5.3 Application of the PAC Model

We now apply the PAC model to analyze the number of examples needed for LEARNPRIM to learn a good selection of primary effects. We first instantiate the PAC model in terms of our primary effect learning problem.

The primary-effect learning problem could be modeled as follows. An example instance is a pair consisting of an initial state and a goal condition. The instance space is the space of all *(Initial, Goal)* pairs. A concept, or a hypothesis, is a selection P of primary effects on operators $\mathcal{O}$. P gives rise to a set of (*init*, *goal*) pairs for which equation (13.1) is satisfied. The hypothesis space consists of all possible selections of primary effects. Let $|E|$ be the total number of operator/effect pairs. The size of the hypothesis space is $|H| = 2^{|E|}$.

A selection P of primary effects is consistent with a problem instance *(Initial, Goal)* if a solution can be found by PRIMPOP such that condition (13.1) is satisfied. Given this mapping, we know that any consistent selection of primary effects satisfies the PAC property as long as the number of training examples is determined by the following formula:

$$\frac{1}{\epsilon}(\ln\frac{1}{\delta} + \ln 2^{|E|}) = \frac{1}{\epsilon}(\ln\frac{1}{\delta} + |E|\ln 2) \tag{13.2}$$

In the above formula, it is assumed that the small positive values ϵ and δ are given. This formula forms a basis for an implementation of the termination condition TERMINATE(), for the LEARNPRIM algorithm; a counter will be incremented until its value exceeds the one in equation (13.2).

What remains to be verified is whether the LEARNPRIM algorithm returns a consistent hypothesis. This condition is needed for the PAC argument to follow. Recall that subroutine UPDATEPRIM converts effects into primary effects based on the causal links in a solution plan Π_p. After this is done, the next time PRIMPOP is called to solve the same planning problem, condition (13.1) must be satisfied, since this time around PRIMPOP will return a plan of the same quality as the optimal solution, Π_o. Applying this argument to every example problem used in learning, we know that when the algorithm terminates, the selected primary effects must be consistent with all examples used in learning. Therefore, when it terminates, the algorithm LEARNPRIM returns a selection of primary effects satisfying the PAC property.

As an example of the learning model, consider a domain with ten operators and two effects for each operator. This domain gives a total of $E = 20$ operator-effect pairs in the domain. If we would like to guarantee that, with a probability higher than 0.95 (that is, $\delta = 0.05$), our selection of primary effects to solve 90% of the planning problems that typically occur, that is, $\epsilon = 0.1$, we could present the learning algorithm with $m \leq 10(\ln(1/0.05) + 20\ln 2)$ examples. In other words, LEARNPRIM should stop when it sees 168 examples.

13.6 Open Problems

Applying PALO to Learning Primary Effects

The PALO algorithm by Cohen, Greiner, and Schuurmans [32] provides an efficient approach to inductively learning theories from given samples. An application of that system to the primary-effect learning problem is an interesting approach. For example, using the PALO system it might be possible for the learning algorithm to terminate earlier than indicated by equation (13.2), which in turn makes it possible to select fewer primary effects than the algorithm we presented.

Problem Distribution

A good selection of primary effects clearly depends on a set of the frequently occurring problems. The problem distribution on this set, however, could change with time. During a regular teaching term a professor might spend more effort solving problems related to teaching and examination marking. In the summer, however, the same person might encounter more travel-related planning problems. The primary-effect learning algorithm works with a fixed, although unknown, distribution of problems. It would be useful for a planner to adjust its primary-effect selection with a changing distribution.

Problem-Dependent Primary Effects

Depending on the particular problem at hand, the same effect of an operator might or might not be considered as a primary effect. Our algorithm LEARNPRIM could be applied to different sets of training examples and therefore result in different selections of primary effects. The context for a given selection is determined by the initial and goal states used in the training process. An interesting research issue is how to compress the different selections of primary effects into a single, compact operator representation. One idea is to combine primary effects with the conditional effects of an extended operator representation, to yield a more effective operator representation.

13.7 Summary

In this chapter we have developed a formal understanding of planning with primary effects and addressed the issue of how primary effects could be selected based on a set of training examples. We pointed out two problems associated with primary effects, namely the loss of completeness for a planner and the possibility of sub-optimality. In addressing the issues, we have introduced a PAC model of primary effects, and based on the model, we described an inductive-learning algorithm for automatically choosing them for improving the solution quality.

13.8 Background

Machine learning and planning have a long history of working together. For a comprehensive overview, see Minton's book [97]. However, few attempts have been made to combine PAC learning and planning, as we have done in this chapter.

The material in this chapter is adapted from two IJCAI papers with Eugene Fink, [47] and [48]. The latter contains extensive experimental results confirming the utility of using primary effects in planning. The learning algorithm and the PAC proof contained in this chapter have been considerably modified to simplify the presentation.

Much of the author's knowledge in PAC learning has been the result of enjoyable discussions with Ming Li. For an introduction to PAC learning theory, see books by Anthony and Biggs [8] and by Kearns and Vazirani [78].

References

1. Philip E. Agre and David Chapman. Pengi: An implementation of a theory of activity. In *Proceedings of the 10th International Joint Conference on Artificial Intelligence (IJCAI-87)*, pages 268–272, Milan, Italy, 1987. Morgan Kaufmann, San Mateo, CA, 1987.
2. Alfred V. Aho, John E. Hopcroft, and Jeffery D. Ullman. *Data Structures and Algorithms*. Addison-Wesley, Reading, MA, 1983.
3. James F. Allen. Towards a general theory of action and time. *Artificial Intelligence*, 23(2):123–154, 1984.
4. James F. Allen. *Natural Language Understanding*. Benjamin/Cummings, Menlo Park, CA, 1987.
5. James F. Allen, James Hendler, and Austin Tate. *Readings in Planning*. Morgan Kaufmann, San Mateo, CA, 1990.
6. James F. Allen, Henry A. Kautz, Richard N. Pelavin, and Josh D. Tenenberg. *Reasoning about Plans*. Morgan Kaufmann, San Mateo, CA, 1991.
7. James F. Allen and Johannes A. Koomen. Planning using a temporal world model. In *Proceedings of the 8th International Joint Conference on Artificial Intelligence (IJCAI-83)*, pages 741–747. Morgan Kaufmann, San Mateo, CA, 1983.
8. Martin Anthony and Norman Biggs. *Computational Learning Theory*. Cambridge Tracts in Theoretical Computer Science 30. Cambridge University Press, Cambridge, UK, 1992.
9. K. B. Athreya and P. E. Ney. *Branching Processes*. Springer-Verlag, New York, 1972.
10. Fahiem Bacchus and Froduald Kabanza. Using temporal logic to control search in a forward-chaining planner. Technical report, University of Waterloo, Waterloo, Ontario, Canada, 1995. Available via the URL ftp://logos.uwaterloo.ca:/pub/tlplan/tlplan.ps.Z.
11. Fahiem Bacchus and Qiang Yang. The downward refinement property. In *Proceedings of the 12th International Joint Conference on Artificial Intelligence (IJCAI-91)*, pages 286–292, Sydney, Australia, August 1991. Morgan Kaufmann, San Mateo, CA, 1991.
12. Fahiem Bacchus and Qiang Yang. The expected value of hierarchical problem-solving. In *Proceedings of the 10th National Conference on Artificial Intelligence (AAAI-92)*, pages 369–374, San Jose, CA. AAAI Press/MIT Press, 1992.
13. Fahiem Bacchus and Qiang Yang. Downward refinement and the efficiency of hierarchical problem solving. *Artificial Intelligence*, 71:43–100, 1994.
14. Christer Backstrom. *Computational Complexity of Reasoning about Plans*. PhD thesis, Linköping University, Linköping, Sweden, 1992.

15. Christer Backstrom. Finding least constrained plans and IMAL parallel executions is harder than we thought. In *Current Trends in AI Planning: EWSP'93–2nd European Workshop on Planning*, pages 46–59, Vadstena, Sweden, 1993. IOS Press, 1993.
16. Christer Backstrom and Peter Jonsson. Planning with abstraction hierarchies can be exponentially less efficient. In *Proceedings of the 14th International Joint Conference on Artificial Intelligence (IJCAI-95)*, Montreal, Quebec, Canada, August 1995. Morgan Kaufmann, San Mateo, CA, 1995.
17. Anthony Barrett and Daniel S. Weld. Partial order planning: Evaluating possible efficiency gains. *Artificial Intelligence*, 67(1):71–112, 1994.
18. Anthony Barrett and Daniel S. Weld. Task-decomposition via plan parsing. In *Proceedings of the 12th National Conference on Artificial Intelligence (AAAI-94)*, pages 1117–1122, Seattle, WA, 1994. AAAI Press/MIT Press, 1994.
19. Fahiem Bascchus and Froduald Kabanza. Planning for temporally extended goals. In *Proceedings of the 13th National Conference on Artificial Intelligence*, volume 2, pages 1214–1222, Menlo Park, CA, August 1996. AAAI Press/MIT Press, 1996.
20. Ralph Bergmann and Wolfgang Wilke. Building and refining abstract planning cases by change of representation language. *Journal of Artificial Intelligence Research*, 3:53–118, 1995. Available electronically at http://www.cs.washington.edu/research/jair/.
21. A. Blum and M.L. Furst. Fast planning through planning graph analysis. In *Proceedings of the 14th International Joint Conference on Artificial Intelligence (IJCAI-95)*, pages 1636–1642, Montreal, Quebec, Canada, 1995, Morgan Kaufmann, San Mateo, CA, 1995.
22. John M. Britanik and Michael M. Marefat. Hierarchical plan merging with applications to process planning. In *Proceedings of the 14th International Joint Conference on Artificial Intelligence (IJCAI-95)*, pages 1677–1685, Montreal, Quebec, Canada, 1995, Morgan Kaufmann, San Mateo, CA, 1995.
23. Rodney A. Brooks. Intelligence without reasoning. In *Proceedings of the 12th International Joint Conference on Artificial Intelligence (IJCAI-91)*, pages 569–595, Sydney, Australia, 1991. Morgan Kaufmann, San Mateo, CA, 1991.
24. Tom Bylander. Complexity results for serial decomposibility. In *Proceedings of the 10th National Conference on Artificial Intelligence (AAAI-92)*, pages 729–734, San Jose, CA, 1992. AAAI Press/MIT Press, 1992.
25. Jaime G. Carbonell, Craig A. Knoblock, and Steven Minton. Prodigy: An integrated architecture for planning and learning. In *Architectures for Intelligence*, pages 241–278. Lawrence Erlbaum, Hillsdale, NJ, 1990.
26. Alex Y.M. Chan. On variable binding commitments in planning. Master's thesis, University of Waterloo, Waterloo, Ontario, Canada 1994.
27. T. C. Chang and R. A. Wysk. *An Introduction to Automated Process Planning Systems.* Prentice-Hall, Englewood Cliffs, NJ, 1985.
28. David Chapman. Planning for conjunctive goals. *Artificial Intelligence*, 32:333–377, 1987.
29. Eugene Charniak and Drew McDermott. *Introduction to Artificial Intelligence.* Addison-Wesley, Reading, MA, 1985.
30. Peter Cheeseman, Bob Kanefsky, and William M. Taylor. Where the really hard problems are. In *Proceedings of the 12th International Joint Conference on Artificial Intelligence (IJCAI-91)*, pages 331–340, Sydney, Australia, August 1991. Morgan Kaufmann, San Mateo, CA, 1991.
31. Jens Christensen. A hierarchical planner that creates its own hierarchies. In *Proceedings of the 8th National Conference on Artificial Intelligence (AAAI-90)*, pages 1004–1009, 1990. AAAI Press/MIT Press, 1990.

32. William W. Cohen, Russ Greiner, and Dale Schuurmans. Probabilistic hill-climbing. In S.J. Hanson, T. Petsche, M. Kearns, and R.L. Rivest, editors, *Computational Learning Theory and Natural Learning Systems*, Vol. 2, pages 171–181. MIT Press, Cambridge, MA, 1992.
33. J.M. Crawford and L.D. Auton. Experimental results on the cross-over point in satisfiability problems. In *Proceedings of the 11th National Conference on Artificial Intelligence (AAAI-93)*, pages 21–27, Washington, DC, 1993. AAAI Press/MIT Press, 1993.
34. Bart Selman David Mitchell, Hector Levesque, and David Mitchell. Hard and easy distributions of sat problems. In *Proceedings of the 10th National Conference on Artificial Intelligence (AAAI-92)*, pages 459–465, San Jose, CA, July 1992. AAAI Press/MIT Press, 1992.
35. Thomas L. Dean and Michael P. Wellman. *Planning and Control.* Morgan Kaufmann, San Mateo, CA, 1991.
36. Rina Dechter. From local to global consistency. *Artificial Intelligence*, 55(1):87–107, 1992.
37. Rina Dechter and Judea Pearl. Network-based heuristics for constraint-satisfaction problems. *Artificial Intelligence*, 34, 1987.
38. Mark Drummond and John Bresina. Anytime synthetic projection: Maximizing the probability of goal satisfaction. In *Proceedings of the 8th National Conference on Artificial Intelligence (AAAI-90)*, pages 138–144, Boston, MA, 1990. AAAI Press/MIT Press, 1990.
39. Mark Drummond and Ken Currie. Exploiting temporal coherence in nonlinear plan construction. *Computational Intelligence*, 4(2):341–348, 1988.
40. Edmond H. Durfee. *Coordination of Distributed Problem Solvers.* Kluwer, Dordrecht, Netherlands, 1988.
41. Edmond H. Durfee and V.R. Lesser. Using partial global plans to coordinate distributed problem solvers. In *Proceedings of the 10th International Joint Conference on Artificial Intelligence (IJCAI-87)*, pages 875–883, 1987. Morgan Kaufmann, San Mateo, CA, 1987.
42. Eithan Ephrati and Jeffrey S. Rosenschein. Divide and conquer in multi-agent planning. In *Proceedings of the 12th National Conference on Artificial Intelligence (AAAI-94)*, pages 375–380, Seattle, WA, 1994. AAAI Press/MIT Press, 1994.
43. Kutluhan Erol, James Hendler, and Dana S. Nau. HTN planning: Complexity and expressivity. In *Proceedings of the 12th National Conference on Artificial Intelligence (AAAI-94)*, pages 1123–1128, Seattle, WA, 1994. AAAI Press/MIT Press, 1994.
44. Oren Etzioni. Acquiring search-control knowledge via static analysis. *Artificial Intelligence*, 62(2):225–302, 1993.
45. Oren Etzioni, Steve Hanks, Daniel Weld, Denise Draper, Neal Lesh, and Mike Williamson. An approach to planning with incomplete information. In *Proceedings of the 3rd International Conference on Principles of Knowledge Representation and Reasoning*, pages 115–125. Morgan Kaufmann, San Mateo, CA, 1992.
46. Richard Fikes and Nils Nilsson. STRIPS: A new approach to the application of theorem proving to problem solving. *Artificial Intelligence*, 2:189–208, 1971.
47. Eugene Fink and Qiang Yang. Characterizing and automatically finding primary effects in planning. In *Proceedings of the 13th International Joint Conference on Artificial Intelligence (IJCAI-93)*, pages 1374–1379, August 1993. Morgan Kaufmann, San Mateo, 1993.

48. Eugene Fink and Qiang Yang. Planning with primary effects: experiments and analysis. In *Proceedings of the 14th International Joint Conference on Artificial Intelligence (IJCAI-95)*, pages 1606–1612, Montreal, Quebec, Canada, 1995. Morgan Kaufmann, San Mateo, CA, 1995.
49. David E. Foulser, Ming Li, and Qiang Yang. Theory and algorithms for plan merging. *Artificial Intelligence*, 57(2):143–182, 1992.
50. Mark S. Fox. *Constraint-Directed Search: A Case Study of Job-Shop Scheduling*. Morgan Kaufmann, San Mateo, CA, 1987.
51. Mark S. Fox, N. Sadeh, and C. Baykan. Constrained heuristic search. In *Proceedings of the 11th International Joint Conference on Artificial Intelligence (IJCAI-89)*, pages 309–315, Detroit, MI, 1989. Morgan Kaufmann, San Mateo, CA, 1989.
52. Mark S. Fox and Monte Zweben. *Intelligent Scheduling*. Morgan Kaufmann, San Mateo, CA, 1995.
53. E.C. Freuder. A sufficient condition of backtrack-free search. *Journal of the ACM*, 29(1):23–32, 1982.
54. Eugene Freuder and Alan Mackworth. Special issue on constraint-directed reasoning. *Artificial Intelligence*, 58(1–3), 1992.
55. Michael R. Garey and David S. Johnson. *Computers and Intractability*. W. H. Freeman and Co., San Francisco, CA, 1979.
56. Malik Ghallab and Herve Laruelle. Representation and control in ixtet, a temporal planner. In Kristian Hammond, editor, *Proceedings of the 2nd International Conference on AI Planning Systems (AIPS-94)*, pages 61–67, Morgan Kaufmann, San Mateo, CA, 1994.
57. M. Ginsberg. Approximate planning. *Artificial Intelligence*, 76, 1995.
58. Fausto Giunchigilia and Toby Walsh. Abstract theorem proving. In *Proceedings of the 11th International Joint Conference on Artificial Intelligence (IJCAI-89)*, pages 372–377, Detroit, MI, 1989. Morgan Kaufmann, San Mateo, CA, 1989.
59. Cordell Green. Application of theorem proving to problem solving. In *Proceedings of the 1st International Joint Conference on Artificial Intelligence (IJCAI-69)*, pages 219–239, 1969. Morgan Kaufmann, San Mateo, CA, 1969.
60. Naresh Gupta and Dana S. Nau. Complexity results in blocks world planning. In *Proceedings of the 9th National Conference on Artificial Intelligence (AAAI-91)*, pages 629–633, Anaheim, CA, July 1991. AAAI Press/MIT Press, 1991.
61. Joseph Y. Halpern and Yorham Moses. A guide to the modal logics of knowledge and belief. In *Proceedings of the 9th International Joint Conference on Artificial Intelligence (IJCAI-85)*, pages 480–490, 1985. Morgan Kaufmann, San Mateo, CA, 1985.
62. R.M. Haralick and G.L. Elliott. Increasing tree-search efficiency for constraint satisfaction problems. *Artificial Intelligence*, 14:263–313, 1980.
63. Caroline C. Hayes. A model of planning for plan efficiency: Taking advantage of operator overlap. In *Proceedings of the 11th International Joint Conference on Artificial Intelligence (IJCAI-89)*, Detroit, MI, 1989. Morgan Kaufmann, San Mateo, CA, 1989.
64. Barbara Hayes-Roth and F. Hayes-Roth. A cognitive model of planning. *Cognitive Science*, 3:275–310, 1979.
65. Joachim Hertzberg and Alexander Horz. Towards a theory of conflict detection and resolution in nonlinear plans. In *Proceedings of the 11th International Joint Conference on Artificial Intelligence (IJCAI-89)*, pages 937–942, Detroit, MI, 1989. Morgan Kaufmann, San Mateo, CA, 1989.

66. Frederick S. Hillier and Gerald J. Lieberman. *Introduction to Operations Research.* McGraw-Hill, New York, 1990.
67. Robert C. Holte, T. Mkadmi, R.M. Zimmer, and A.J. MacDonald. Speeding up problem solving by abstraction: A graph oriented approach. *Artificial Intelligence*, 85(1–2):321–361, 1996.
68. David Joslin and Martha E. Pollack. Least-cost flaw repair: A plan refinement strategy for partial-order planning. In *Proceedings of the Twelfth National Conference on Artificial Intelligence*, Vol. 2, pages 1004–1009, Menlo Park, CA, July 1994. AAAI Press/MIT Press, 1994.
69. David Joslin and Martha E. Pollack. Passive and active decision postponement in plan generation. In M. Ghallab and A. Milani, editors, *New Directions in AI Planning*, pages 37–48. IOS Press, Amsterdam, Netherlands, 1997.
70. Subbarao Kambhampati. *Flexible Reuse and Modification in Hierarchical Planning: A Validation Structure Based Approach.* PhD thesis, University of Maryland, College Park, MD, Oct. 1989.
71. Subbarao Kambhampati. Characterizing multi-contributor causal structures for planning. In *Proceedings of the 1st International Conference on AI Planning Systems (AIPS-92)*, pages 116–125. Morgan Kaufmann, San Mateo, CA, 1992.
72. Subbarao Kambhampati. On the utility of systematicity: understanding tradeoffs between redundancy and commitment in partial-order planning. In *Proceedings of the 13th International Joint Conference on Artificial Intelligence (IJCAI-93)*, pages 1380–1385, Chambery, Savoie, France, August 1993. Morgan Kaufmann, San Mateo, CA, 1993.
73. Subbarao Kambhampati and James A. Hendler. A validation structure based on theory of plan modification and reuse. *Artificial Intelligence*, 55(2–3), 1992.
74. Subbarao Kambhampati, Craig A. Knoblock, and Qiang Yang. Planning as refinement search: A unified framework for evaluating design tradeoffs in partial-order planning. *Artificial Intelligence*, 75(3):167–238, 1995. Special Issue on Planning and Scheduling, edited by J. Hendler and D. McDermott.
75. Subbarao Kambhampati and Dana S. Nau. On the nature of modal truth criterion in planning. In *Proceedings of the 12th National Conference on Artificial Intelligence (AAAI-94)*, pages 1055–1060. AAAI Press/MIT Press, 1994.
76. Raghu Karinthi, Dana S. Nau, and Qiang Yang. Handling feature interactions in process-planning. *Applied Artificial Intelligence*, 6(4):389–415. 1992.
77. Henry Kautz and Bart Selman. Pushing the envelope: Planning, propositional logic and stochastic search. In *Proceedings of the 13th National Conference on Artificial Intelligence*, Vol. 2, pages 1194–1201, Menlo Park, CA, August 1996. AAAI Press/MIT Press, 1996.
78. Michael J. Kearns and Umesh V. Vazirani. *An Introduction to Computational Learning Theory.* MIT Press, Cambridge, MA, 1994.
79. Craig A. Knoblock, Josh Tenenberg, and Qiang Yang. Characterizing abstraction hierarchies for planning. In *Proceedings of the 9th National Conference on Artificial Intelligence (AAAI-91)*, pages 692–696, Anaheim, CA, 1991. AAAI Press/MIT Press, 1991.
80. Craig A. Knoblock. *Generating Abstraction Hierarchies — An automated approach to reducing search in planning.* Kluwer, Dordrecht, Netherlands, 1993.
81. Craig A. Knoblock and Qiang Yang. Evaluating the tradeoffs in partial-order planning algorithms. In *Proceedings of the Canadian Artificial Intelligence Conference, (AI 94)*, pages 279–286, 1994. Winner of best paper award. Canadian Society for Computational Studies of Intelligence, 1994.
82. Janet Kolodner. *Case-Based Reasoning.* Morgan Kaufmann, San Mateo, CA, 1993.

83. Richard Korf. Planning as search: A quantitative approach. *Artificial Intelligence*, 33:65–88, 1985.
84. Vipin Kumar. Algorithms for constraint satisfaction problems: A survey. *AI Magazine*, Spring, 32–44, 1992.
85. Nicholas Kushmerick, Steve Hanks, and Daniel Weld. An algorithm for probabilistic least-commitment planning. In *Proceedings of the 12th National Conference on Artificial Intelligence (AAAI-94)*, pages 1073–1078, Seattle, WA, 1994. AAAI Press/MIT Press, 1994.
86. Pat Langley. *Elements of Machine Learning.* Morgan Kaufmann, San Francisco, CA, 1995.
87. Amy L. Lansky and Lise Getoor. Scope and abstraction: Two criteria for localized planning. In *Proceedings of the 14th International Joint Conference on Artificial Intelligence (IJCAI-95)*, pages 1612–1619, Montreal, Quebec, Canada, 1995. Morgan Kaufmann, San Mateo, CA, 1995.
88. Amy L. Lansky. Localized event-based reasoning for multiagent domains. *Computational Intelligence*, 4(4):319–340, 1988.
89. Amy L. Lansky. Localized planning with diversified plan construction methods. Technical Report FIA-93-17, NASA Ames Research Center, AI Research Branch, 1993.
90. Vladimir Lifschitz. On the semantics of STRIPS. In Michael Geogeff and Amy Lansky, *Reasoning about Plans*, Morgan Kaufmann, San Mateo, CA, 1997.
91. Alan K. Mackworth. Consistency in networks of relations. In Webber and Nilsson, editors, *Readings in Artificial Intelligence*, pages 69–78. Morgan Kaufmann, San Mateo, CA, 1981.
92. Alan K. Mackworth and Eugene C. Freuder. The complexity of some polynomial network consistency algorithms for constraint satisfaction problems. *Artificial Intelligence*, 125:65–74, 1985.
93. M. Mantyla and J. Opas. Hutcapp—a machining operations planner. In *Proceedings of the Second International Symposium on Robotics and Manufacturing systems*, 1988.
94. David McAllester and David Rosenblitt. Systematic nonlinear planning. In *Proceedings of the 9th National Conference on Artificial Intelligence (AAAI-91)*, pages 634–639, AAAI Press/MIT Press, 1991.
95. Steve Minton, John Bresina, and Mark Drummond. Commitment strategies in planning: A comparative analysis. In *Proceedings of the 12th International Joint Conference on Artificial Intelligence (IJCAI-91)*, pages 259–265, Sydney, Australia, 1991. Morgan Kaufmann, San Mateo, CA, 1991.
96. Steve Minton, Mark D. Johnston, Andrew B. Philips, and Philip Laird. Solving large scale constraint satisfaction and scheduling problems using a heuristic repair method. In *Proceedings of the 8th National Conference on Artificial Intelligence (AAAI-90)*, pages 17–24, Boston, 1990. AAAI Press/MIT Press, 1990.
97. Steven Minton, editor. *Machine Learning Methods for Planning.* Morgan Kaufmann, San Mateo, CA, 1993.
98. Dana S. Nau. Automated process planning using hierarchical abstraction. *Texas Instruments Technical Journal*, pages 39–46, Winter 1987. Award winner, Texas Instruments 1987 Call for Papers on Industrial Automation.
99. Dana S. Nau, Vipin Kumar, and Laveen Kanal. General branch and bound, and its relation to A* and Ao*. *Artificial Intelligence*, 23:29–58, 1984.
100. Dana S. Nau, Qiang Yang, and James Hendler. Optimization of multiple-goal plans with limited interaction. In *Proceedings of the 1990 DARPA Workshop*

on Innovative Approaches to Planning, Scheduling and Control, pages 160–165, San Diego, CA, November 1990. Morgan Kaufmann, San Mateo, CA, 1990.

101. Allen Newell and Herbert A. Simon. *Human Problem Solving.* Prentice-Hall, Englewood Cliffs, NJ, 1972.
102. Nils Nilsson. *Principles of Artificial Intelligence.* Morgan Kaufmann, San Mateo, CA, 1980.
103. Edwin P.D. Pednault. *Towards a mathematical theory of plan synthesis.* PhD thesis, Stanford University, December 1986.
104. Edwin P.D. Pednault. Synthesizing plans that contain actions with context-dependent effects. *Computational Intelligence*, 4(4):356–372, 1988.
105. J. Scott Penberthy and Daniel S. Weld. UCPOP: A sound, complete, partial order planner for ADL. In *Proceedings of the 1992 International Conference on Principles of Knowledge Representation and Reasoning*, pages 103–114. Morgan Kaufmann, Los Altos, CA, 1992.
106. Mark A. Peot and David E. Smith. Conditional nonlinear planning. In James Hendler, editor, *Proceedings of the First International Conference on AI Planning Systems (AIPS-92)*, pages 189–197. Morgan Kaufmann, San Mateo, CA, 1992.
107. Mark A. Peot and David E. Smith. Threat-removal strategies for partial-order planning. In *Proceedings of the 11th National Conference on Artificial Intelligence (AAAI-93)*, pages 492–499, Washington, DC, 1993. AAAI Press/MIT Press, 1993.
108. Martha E. Pollack. The uses of plans. *Artificial Intelligence*, 57(1):43–68, 1992.
109. Raymond Reiter. Proving properties of states in the situation calculus. *Artificial Intelligence*, 64(2):337–352, 1993.
110. Elaine Rich and Kevin Knight. *Artificial Intelligence.* McGraw-Hill, New York, 1991.
111. David A. Rosenblitt. *Supporting Collaborative Planning: The Plan Integration Problem.* PhD thesis, MIT, Cambridge, MA, Feb. 1991.
112. Stuart Russell and Peter Norvig. *Artificial Intelligence: A Modern Approach.* Prentice-Hall, Englewood Cliffs, NJ, 1995.
113. Earl Sacerdoti. Planning in a hierarchy of abstraction spaces. *Artificial Intelligence*, 5:115–135, 1974.
114. Earl Sacerdoti. *A Structure for Plans and Behavior.* American Elsevier, New York, 1977.
115. Marcel Schoppers. Universal plans for reactive robots in unpredictable environments. In *Proceedings of the 10th International Joint Conference on Artificial Intelligence (IJCAI-87)*, pages 1039–1046, Milan, Italy, 1987. Morgan Kaufmann, San Mateo, CA, 1987.
116. Len Schubert and A. Gerevini. Accelerating partial-order planners by improving plan and goal choices. Technical Report 96-607, University of Rochester, Department of Computer Science, 1996.
117. Timos Sellis. Multiple query Optimization. *ACM Transactions on Database Systems*, 13(1), March 1988.
118. Bart Selman and Henry Kautz. Domain-independent extensions to GSAT: Solving large structured satisfiability problems. In *Proceedings of the 13th International Joint Conference on Artificial Intelligence (IJCAI-93)*, pages 290–295, Chambery, Savoie, France, August 1993. Morgan Kaufmann, San Mateo, CA, 1993.

119. Bart Selman, Henry Kautz, and Bram Cohen. Noise strategies for improving local search. In *Proceedings of the 12th National Conference on Artificial Intelligence (AAAI-94)*, pages 337–343, Seattle, WA, 1994. AAAI Press/MIT Press, 1994.
120. Bart Selman, Hector Levesque, and David Mitchell. A new method for solving hard satisfiability problems. In *Proceedings of the 10th National Conference on Artificial Intelligence (AAAI-92)*, pages 440–446, San Jose, CA, July 1992. AAAI Press/MIT Press, 1992.
121. L. Siklossy and J. Dreussi. An efficient robot planner which generates its own procedures. In *Proceedings of the 3rd International Joint Conference on Artificial Intelligence (IJCAI-73)*, pages 423–430, 1973. Morgan Kaufmann, San Mateo, CA, 1973.
122. Herbert A. Simon. *Models of Bounded Rationality*, Vol. 2. MIT Press, Cambridge, MA, 1982.
123. David E. Smith and Mark A. Peot. A critical look at Knoblock's hierarchy mechanism. In *Proceedings of the 1st International Conference on AI Planning Systems (AIPS-92)*, pages 307–308, College Park, MD, 1992. AAAI Press, 1992.
124. David E. Smith and Mark A. Peot. Postponing threats in partial-order planning. In *Proceedings of the 11th National Conference on Artificial Intelligence (AAAI-93)*, pages 500–506, Washington, DC, 1993. AAAI Press/MIT Press, 1993.
125. Reid G. Smith and Davis Randall. Frameworks for cooperation in distributed problem solving. *IEEE Transactions on Systems, Man and Cybernetics*, 11(1):61–70, 1981.
126. R. Sosic and J. Gu. A polynomial time algorithm for the n-queens problem. In *SIGART 1(3)*, 1990.
127. R. Srinivasan and A. E. Howe. Comparison of methods for improving search efficiency in a partial-order planner. In *Proceedings of the 14th International Joint Conference on Artificial Intelligence (IJCAI-95)*, pages 1620–162, Montreal, Quebec, Canada, 1995. Morgan Kaufmann, San Mateo, CA, 1995.
128. Mark Stefik. Planning with constraints. *Artificial Intelligence*, 16(2):111–140, 1981.
129. Peter Stone, Manuela Veloso, and Jim Blythe. The need for different domain-independent heuristics. In Kristian Hammond, editor, *Proceedings of the Second International Conference on AI Planning Systems (AIPS-94)*, pages 164–169 Morgan Kaufmann, San Mateo, CA, 1994.
130. Gerald J. Sussman. *A Computational Model of Skill Acquisition*. MIT AI Lab Memo AI-TR-297, 1973.
131. Steven L. Tanimoto. *The Elements of Artificial Intelligence Using Common Lisp*. 2nd edn. Computer Science Press, 1995.
132. Austin Tate. Generating project networks. In *Proceedings of the 5th International Joint Conference on Artificial Intelligence (IJCAI-77)*, pages 888–893. Morgan Kaufmann, San Mateo, CA, 1977.
133. Austin Tate, Brian Drabble, and Jeff Dalton. The use of condition types to restrict search in an AI planner. In *Proceedings of the 12th National Conference on Artificial Intelligence (AAAI-94)*, pages 1129–1134, Seattle, WA, 1994. AAAI Press/MIT Press, 1994.
134. Josh Tenenberg. *Abstraction in Planning*. PhD thesis, University of Rochester, Dept. of Computer Science, Rochester, NY, May 1988.
135. James Allen Thomas Dean and Yiannis Aloimonos. *Artificial Intelligence: Theory and Practice*. Benjamin/Cummings Pub. Co., Redwood City, CA, 1995.

136. Scott Thompson. Environment for hierarchical abstraction: A user guide. Master's thesis, Computer Science Department, University of Maryland, College Park, MD, 1989.
137. Sun Tzu. *The Art of Strategy (The Art of War)*. Doubleday, New York, 1988. Translated by R.L. Wing.
138. Amy Unruh and Paul S. Rosenbloom. Abstraction in problem solving and learning. In *Proceedings of the 11th International Joint Conference on Artificial Intelligence (IJCAI-89)*, pages 681–687, Detroit, MI, 1989. Morgan Kaufmann, San Mateo, CA, 1989.
139. Manuela Veloso and Jim Blythe. Linkability: Examining causal link commitments in partial-order planning. In Kristian Hammond, editor, *Proceedings of the 2nd International Conference on AI Planning Systems (AIPS-94)*, pages 170–175. Morgan Kaufmann, San Mateo, CA, 1994.
140. Steven A. Vere. Planning in time: Windows and durations for activities and goals. *IEEE Transactions on Pattern Analysis and Machine Intelligence*, 5:246–267, 1983.
141. Mark Vilain and Henry Kautz. Constraint propagation algorithms for temporal reasoning. In *Proceedings of the 5th National Conference on Artificial Intelligence (AAAI-86)*, pages 337–382, 1986. AAAI Press/MIT Press, 1986.
142. Richard Waldinger. Achieving several goals simultaneously. In Elcock and Michie, editors, *Machine Intelligence 8*. Ellis Horwood, Chichester, UK, 1977.
143. David H.D. Warren. Warplan: A system for generating plans. Memo No. 76, Department of Computational Logic, University of Edinburgh, 1974.
144. Daniel Weld. An introduction to least-commitment planning. *AI Magazine*, Winter, 27–61, 1994.
145. Robert Wilensky. *Planning and Understanding*. Addison-Wesley, Reading, MA, 1983.
146. David Wilkins. Domain-independent planning: Representation and plan generation. *Artificial Intelligence*, 22:269–301, 1984.
147. David E. Wilkins. *Practical Planning: Extending the Classical AI Planning Paradigm*. Morgan Kaufmann, San Mateo, CA, 1988.
148. Patrick H. Winston. *Artificial Intelligence*. 3rd edn. Addison-Wesley, Reading, MA, 1992.
149. Steven G. Woods. An implementation and evaluation of a hierarchical nonlinear planner. Master's thesis, Computer Science Department, University of Waterloo, Ontario, Canada, 1991.
150. Qiang Yang. *Improving the Efficiency of Planning*. PhD thesis, University of Maryland, College Park, MD, July 1989.
151. Qiang Yang. An algebraic approach to conflict resolution in planning. In *Proceedings of the 8th National Conference on Artificial Intelligence (AAAI-90)*, pages 40–45, AAAI Press/MIT Press 1990.
152. Qiang Yang. Formalizing planning knowledge for hierarchical planning. *Computational Intelligence*, 6:12–24, 1990.
153. Qiang Yang. Reasoning about conflicts in least-commitment planning. Technical Report cs-90-23, Department of Computer Science, University of Waterloo, Ontario, Canada, 1990.
154. Qiang Yang. A theory of conflict resolution in planning. *Artificial Intelligence*, 58(1–3), 1992.
155. Qiang Yang and Alex Y.M. Chan. Delaying variable binding commitments in planning. In Kristian Hammond, editor, *Proceedings of the Second International Conference on AI Planning Systems (AIPS-94)*, pages 182–187. Morgan Kaufmann, San Mateo, CA, 1994.

156. Qiang Yang and Philip Fong. Situated control rules as approximate plans: A pac-based theory. In M. Ghallab and A. Milani, editors, *New Directions in AI Planning*, Vol. 1, pages 327–340. IOS Press, September 1996.
157. Qiang Yang and Cheryl Murray. An evaluation of the temporal coherence heuristic in partial-order planning. *Computational Intelligence*, 10(3):245–267, 1994.
158. Qiang Yang, Dana S. Nau, and James Hendler. Merging separately generated plans in limited domains. *Computational Intelligence*, 8(4):648–676, 1992.
159. Qiang Yang and Josh Tenenberg. ABTWEAK: Abstracting a nonlinear, least commitment planner. In *Proceedings of the 8th National Conference on Artificial Intelligence (AAAI-90)*, pages 204–209, Boston, MA, August 1990. AAAI Press, 1990.
160. Qiang Yang, Josh Tenenberg, and Steve Woods. On the implementation and evaluation of AVTWEAK. *Computational Intelligence*, 12:307–330, 1996.
161. Michael Young, Martha E. Pollack, and Johanna D. Moore. Decomposition and causality in partial-order planning. In K. Hammond, editor, *Proceedings of the 2nd International Conference on AI Planning Systems (AIPS-94)*, pages 188–193, Chicago, IL, 1994. Morgan Kaufmann, San Mateo, CA, 1994.
162. Edward Tsang. *Foundations of Constraint Satisfaction.* Academic Press, London, UK, 1993.
163. Kristian Hammond. Explaining and repairing plans that fail. *Artificial Intelligence Journal* 45(1):173–228, 1990.
164. Eugine Fink and Quang Yang. Automatically selecting and using primary effects in planning: Theory and experiments. *Artificial Intelligence Journal* 89(1–2):285–316, 1997.

Index

abstract operators 166
abstraction
– introduction 3, 161
Agre, P. 9
Aho, A. 66
Aho, A., Hopcroft, J. and Ullman, J. 8
algorithm
– basic planners
– – critics 46
– branch and bound 56
– – listing 57
– conversion from total order to partial order 28
– CSP 58
– – CspSolve 77
– – arc-consistency 59
– – backtrack free 60
– – backtracking 59
– – consistency methods 59
– – local search 59
– – value-ordering heuristics 62
– dynamic programming 54
– – in plan merging 132
– – listing 55
– graph
– – connectivity 51
– – cycle detection 51
– – strongly connected components 52
– – topological sort 51
– inductive learning 235
– planning
– – Abstrips 96, 163
– – AbTweak 37, 163, 226
– – Alpine 51, 100, 188
– – Dcomp 100
– – FdPop 71
– – Gemplan 79
– – GPS 163
– – GraphPlan 9, 38
– – Gsat 66, 79
– – Hipoint 51, 192
– – IxTeT 79
– – Lawly 163
– – Loc 99
– – Machinist 137
– – Molgen 78, 163
– – Noah 163
– – Poplan 78
– – Prodigy 120, 226
– – SatPlan 9, 35, 37, 38
– – Sipe 163, 188, 226
– – Snlp 37, 48
– – Soar 163
– – Static 120
– – Strips 100, 226, 229
– – TlPlan 38
– – ToPlan 28
– – Tweak 100
– – Ucpop 29
– – Watplan 143
– – approximate solution 25
– – backward-chaining 20
– – basic planners 19
– – forward chaining 27
– unification 50
– – listing 51
– – substitution 49
– – unifier 50
– – variable binding 50
Allen, J. 8, 79, 188
Aloimonos, Y. 8
analysis
– AbTweak 44
– Tweak 44, 46
– forward-chaining, total-order planners 40
– partial-order planners 41
– abstraction 179
Anthony, M. 237
assumptions in abstraction 181
Auton, L. D. 37

Bacchus 38, 48, 188
Backstrom 8, 28, 187, 204
Barrett 8, 37, 48, 100, 223
basic planner
– discussion 20
– introduction 4
Bergmann, R. 187, 206
Biggs, N. 237
Blythe, J. 37
Bresina, J. 8, 48
Brooks, R. 9
Bylander, T. 8

Carbonell, J. 120
causal links
– definition 19
Chan, A. 79
Chapman, D. 8, 9, 100, 120
Charniak, E. 8
Cheeseman, P. 66, 186
commitment strategy
– delayed-commitment 67
– eager commitment 69
– least-commitment 67
complexity of planning 8
conflict resolution 101
– augmented arc-consistency 109
– forward checking 110
– in HACKER 115
– in MOLGEN 115
– in NOAH 115
– in NONLIN 115
– in SIPE 115
– in TWEAK 115
– in social/political situations 119
– inconsistency relation 104
– minimal solution 106
– redundancy removal 108
– represented as constraint satisfaction 107
– subsumption relation 105
– unresolvable conflicts 217
– work by Smith and Peot 116
contract net 99
CPM
– critical path method 5
Crawford, J.M. 37
criticality 166
CSP 58, 67, 77
Currie, K. 8, 48

Davis–Putnam procedure 37
Dean, T. 8
Dechter, R. 66
decision theory 9
decomposition 3
– benefits 90, 91
– in STRIPS 97
– in abstraction 96
– in partial-order planners 98
– localization 98
Drummond, M. 8, 48
Durfee, E. 9

Ephrati, E. 120, 158
Erol, K. 223
Etzioni, O. 48, 118, 120

Fikes, R. 8
finite resources 71
Fink, E. 237
Fong, P.W. 25
Foulser, D. 139
Fox, M. 9, 78
Freuder, E. 66, 117

Getoor, L. 206
Ghallab, M. 79
Ginsberg, M. 25
Giunchiglia, F. 188
Green, C. 8
Gupta, N. 8

Halpern, J. 9
Haralick, R. 66
Hart, P. 8
Hayes, P. 123, 134
Hendler, J. 8, 137, 139, 158, 223
Hertzberg, J. 120
hierarchical planning 165
Holte, R. 188
Hopcroft, J. 66
Horz, A. 120
HTN planning 207

inductive learning algorithm, 235

Jonsson, P. 187, 204
Joslin 37, 48, 79, 115

Kabanza, F. 38, 48
Kambhampati 8, 37, 48, 137
Karinthi, R. 158
Kautz, H. 9, 37, 38, 66, 79, 118, 188
Kearns, M. 237
Knight, K. 8, 66
Knoblock, C. 8, 48, 188, 193, 206

Kolodner, J. 9
Korf, R. 48, 188
Kumar, V. 66

Lansky, A. 8, 79, 206
Levesque, H. 37
Li, M. 139, 237
Lieberman 9
Lifschitz, V. 9
local-search methods 9, 38, 66, 79

machine learning 237
– for primary effects 235
Mackworth, A. 66
McAllester, D. 8
McDermott, D. 8
Minton, S. 8, 48, 237
Mitchell, T. 37
multi-agent planning 9

Nau, D.S. 8, 139, 158, 223
Newell, A. 8
Nilsson, N. 8, 66
NOAH 8
NONLIN 8
Norvig, P. 66

office building construction 8
open precondition
– definition 22
operator overloading 138
operator reduction 208
operator representation
– advanced 29
– costs 16
– effects 16
– preconditions 16
– conditional effects 30

PAC learning 233, 234
partial order
– definition 18
– introduction 5
partial-order plan
– definition 18
Pednault, E. 8, 38
Peot, M. 48, 206
plan merging 3
– approximate algorithm 134
– blocks-world example 125
– branch and bound algorithm 149
– complexity 129, 148
– definition 123, 124
– grouping operators 124
– hierarchical 139
– in Machinist 123
– upper bound 152
– with conflict resolution 139
– worst case analysis 135
planning domains
– TOH(Towers of Hanoi) 168
– painting domain 16
– travel 170
– variable-sized blocks world 76
Pollack, M. 37, 48, 79, 115, 138
primary effects
– impact on completeness 230
– in ABTWEAK 226
– in PRODIGY 226
– in SIPE 226
– in STRIPS 226
– introduction 4, 226
– use in depth-first search 229
primary-effect restricted planner 231
Probably Approximately Correct learning 234
project management 9
properties
– abstraction hierarchy
– – downward refinement 178
– – downward solution 175
– – monotonic 176
– – ordered monotonic 177
– – syntactic connectivity conditions 189
– – upward solution 174
– admissible 24
– completeness 24
– correctness 21

QA3 8

refinement
– length-first 172
– partial-order 172
reflexive relations 104
Rich, C. 8, 66
Rosenblitt, D. 8
Rosenschein, J. 120, 158
Russell, S. 66

Sacerdoti, E. 8, 123, 134, 166, 188, 223
Sadeh, N. 78
scheduling 4
– interleave with planning 6
Schoppers, M. 9
SCS (Shortest Common Supersequence) 130

Selman, B. 9, 37, 38, 66, 79
Simon, H. 8, 138
Smith, D. 48, 206
soundness and completeness
– problems associated with 24
spacecraft planning 8
step orders
– definition 18
Stone, P. 37
Sussman, G. 8, 120

Tanimoto, S. 8
Tate, A. 8, 120, 123, 223
Taylor 66
temporal operator language 118
Tenenberg, J. 8, 37, 188
threat
– negative 22
– positive 22
threat resolution 101
– confrontation 30
– demotion 23
– promotion 23
– separation 23
– unresolvable conflicts 217
– white knights 37, 44
total-order plan
– definition 18

Ullman, J. 66
unifiable
– definition 22
unification 24
unique-main-subaction restriction 217
Unruh, A. 163
utility problem
– in conflict resolution 112
– in finite-resource planning 67

variable-binding constraints
– definition 18, 22
– instantiation 16
Veloso, M. 37
Vere, S. 8
Vilain, M. 118

Waldinger, R. 100, 120
Walksat 37
Walsh, T. 188
Warren, D. 8
Weld, D. 8, 37, 48, 100, 223
Wellman, M. 9
Wilensky, R. 9, 137
Wilke, W. 187, 206
Wilkins, D. 8, 120, 123, 134, 137, 188, 223
Winston, P. 8
Woods, S. 37, 188, 206

Yang, Q. 8, 25, 37, 48, 79, 120, 139, 158, 188, 223, 237
Young, M. 223

Zweben, M. 9, 78

Springer Series
Artificial Intelligence

N. J. Nilsson: Principles of Artificial Intelligence. XV, 476 pages, 139 figs., 1982

J. H. Siekmann, G.Wrightson (Eds.): Automation of Reasoning 1. Classical Papers on Computational Logic 1957–1966. XXII, 525 pages, 1983

J. H. Siekmann, G. Wrightson (Eds.): Automation of Reasoning 2. Classical Papers on Computational Logic 1967–1970. XXII, 638 pages, 1983

L. Bolc (Ed.): The Design of Interpreters, Compilers, and Editors for Augmented Transition Networks. XI, 214 pages, 72 figs., 1983

M. M. Botvinnik: Computers in Chess. Solving Inexact Search Problems. With contributions by A. I. Reznitsky, B. M. Stilman, M. A. Tsfasman, A. D.Yudin. Translated from the Russian by A. A. Brown. XIV, 158 pages, 48 figs., 1984

L. Bolc (Ed.): Natural Language Communication with Pictorial Information Systems. VII, 327 pages, 67 figs., 1984

R. S. Michalski, J. G. Carbonell, T. M. Mitchell (Eds.): Machine Learning. An Artificial Intelligence Approach. XI, 572 pages, 1984

C. Blume, W. Jakob: Programming Languages for Industrial Robots. XIII, 376 pages, 145 figs., 1986

J. W. Lloyd: Foundations of Logic Programming. Second, extended edition. XII, 212 pages, 1987

L. Bolc (Ed.): Computational Models of Learning. IX, 208 pages, 34 figs., 1987

L. Bolc (Ed.): Natural Language Parsing Systems. XVIII, 367 pages, 151 figs., 1987

N. Cercone, G. McCalla (Eds.): The Knowledge Frontier. Essays in the Representation of Knowledge. XXXV, 512 pages, 93 figs., 1987

G. Rayna: REDUCE. Software for Algebraic Computation. IX, 329 pages, 1987

D. D. McDonald, L. Bolc (Eds.): Natural Language Generation Systems. XI, 389 pages, 84 figs., 1988

L. Bolc, M. J. Coombs (Eds.): Expert System Applications. IX, 471 pages, 84 figs., 1988

C.-H. Tzeng: A Theory of Heuristic Information in Game-Tree Search. X, 107 pages, 22 figs., 1988

H. Coelho, J. C. Cotta: Prolog by Example. How to Learn, Teach and Use It. X, 382 pages, 68 figs., 1988

Springer Series
Artificial Intelligence

L. Kanal, V. Kumar (Eds.): Search in Artificial Intelligence. X, 482 pages, 67 figs., 1988

H. Abramson, V. Dahl: Logic Grammars. XIV, 234 pages, 40 figs., 1989

R. Hausser: Computation of Language. An Essay on Syntax, Semantics, and Pragmatics in Natural Man-Machine Communication. XVI, 425 pages, 1989

P. Besnard: An Introduction to Default Logic. XI, 201 pages, 1989

A. Kobsa, W. Wahlster (Eds.): User Models in Dialog Systems. XI, 471 pages, 113 figs., 1989

B. D'Ambrosio: Qualitative Process Theory Using Linguistic Variables. X, 156 pages, 22 figs., 1989

V. Kumar, P. S. Gopalakrishnan, L. N. Kanal (Eds.) Parallel Algorithms for Machine Intelligence and Vision. XI, 433 pages, 148 figs., 1990

Y. Peng, J. A. Reggia: Abductive Inference Models for Diagnostic Problem-Solving. XII, 284 pages, 25 figs., 1990

A. Bundy (Ed.): Catalogue of Artificial Intelligence Techniques. Third, revised edition. XV, 179 pages, 1990

D. Navinchandra: Exploration and Innovation in Design. XI, 196 pages, 51 figs., 1991

R. Kruse, E. Schwecke, J. Heinsohn: Uncertainty and Vagueness in Knowledge Based Systems. Numerical Methods. XI, 491 pages, 59 figs., 1991

Z. Michalewicz: Genetic Algorithms + Data Structures = Evolution Programs. XVII, 250 pages, 48 figs., 1992

V. W. Marek, M. Truszczyński: Nonmonotonic Logic. Context-Dependent Reasoning. XIII, 417 pages, 14 figs., 1993

V. S. Subrahmanian, S. Jajodia (Eds.): Multimedia Database Systems. XVI, 323 pages, 104 figs., 1996

Q. Yang: Intelligent Planning. XXII, 252 pages, 76 figs., 1997

Printing: Mercedesdruck, Berlin
Binding: Buchbinderei Lüderitz & Bauer, Berlin